A HISTORY OF THE IDEAS OF THEORETICAL PHYSICS

Essays on the Nineteenth and Twentieth Century Physics

by

SALVO D'AGOSTINO
Università "La Sapienza",
Roma

KLUWER ACADEMIC PUBLISHERS
DORDRECHT / BOSTON / LONDON

Library of Congress Cataloging-in-Publication Data

D'Agostino, Salvo.
　A history of the ideas of theoretical physics : essays on the nineteenth and twentieth century physics / by Salvo D'Agostino.
　　p. cm. -- (Boston studies in the philosophy of science ; v. 213)
　Includes bibliographical references and index.
　ISBN 0-7923-6094-X (alk. paper)
　1. Physics--History--19th century. 2. Physics--History--20th century. I. Title. II. Series.

Q174 .B67 no. 213
[QC7]
001.'01 s--dc21
[530.1'09'034]
　　　　　　　　　　　　　　　　　　　　　　　　　　　　　　　99-058405
ISBN 0-7923-6094-X

Published by Kluwer Academic Publishers,
P.O. Box 17, 3300 AA Dordrecht, The Netherlands.

Sold and distributed in North, Central and South America
by Kluwer Academic Publishers,
101 Philip Drive, Norwell, MA 02061, U.S.A.

In all other countries, sold and distributed
by Kluwer Academic Publishers,
P.O. Box 322, 3300 AH Dordrecht, The Netherlands.

Printed on acid-free paper

All Rights Reserved
© 2000 Kluwer Academic Publishers
No part of the material protected by this copyright notice may be reproduced or
utilized in any form or by any means, electronic or mechanical,
including photocopying, recording or by any information storage and
retrieval system, without written permission from the copyright owner.

Printed in the Netherlands.

TABLE OF CONTENTS

INTRODUCTION xi

PART ONE: FROM MECHANICS TO ELECTRO-DYNAMICS

Foreword to Part One 3

1. A CONSIDERATION ON THE CHANGING ROLE OF MATHEMATICS IN AMPÈRE'S AND WEBER'S ELECTRODYNAMICS 7

1.1 Ampère's Electrodynamics. 7
1.2 Ampère's Theory of Equivalence and its Foundation on the Symmetries of Magnetic Forces 12
1.3 Ampère's Program: Only Mathematical Laws are Perennially Valid Because They are Independent from Physical Hypotheses 15
1.4 Gauss's and Weber's Metrology. Weber's Discovery of a Constant Velocity of Motion of Electric "Masses". 17
1.5 Comments 34

2. A SURVEY OF THEORIES OF UNITS AND DIMENSIONS IN NINETEENTH-CENTURYPHYSICS 39

3. A HISTORICAL ROLE FOR DIMENSIONAL ANALYSIS IN MAXWELL'S ELECTROMAGNETIC THEORY OF LIGHT 45

3.1 Maxwell Transforms Electrodynamics into Electromagnetism 45
3.2 In Order to Adopt Weber's Result, Maxwell Resorts to

v

the Elastic Theories of Optics 48
3.3 Metrology Makes its Entry on Maxwell's Optical Stage 52
3.4 The Transitional Principle 54
3.5 Elimination of Hypotheses and Experimental
Evidence in a "Dynamical Theory" 56
3.6 Velocity of Light and Comparison of Forces in 1868 58
3.7 Maxwell's Final Achievement: Metrology and
Theory of Dimensions in *A Treatise* 61
3.8 Comments 70

4. PROBLEMS OF THEORETICAL PHYSICS IN THE SECOND HALF OF THE NINETEENTH CENTURY 77

4.1 A Consideration of Maxwell's Innovative Ideas about
Physical Theories 77
4.2 Helmholtz's Secularisation of Kant's A-Priori 87
4.3 Theory and Experiment in Hertz's Holistic
Conception of Theoretical Physics 93
4.4 Mach's Descriptive Ideal and the Elimination of Models .. 102

PART TWO: ELECTROMAGNETIC WAVES

Foreword to Part Two 109

5. GERMAN ELECTRODYNAMICS IN THE 1870'S 111

5.1 Circuital Theories and Faraday's Induction 111
5.2 Prior to Hertz: Experiment on Distant-Induction, not a
Test of Maxwell's Electromagnetic Theory 116
5.3 Hertz's Conciliation of Two Distinct Research
Traditions: Electric Waves in Wires and in Space 119
5.4 Helmholtz's Electrodynamics 123
5.5 Helmholtz's Theory of Dielectric Polarisation 127
5.6 Helmholtz's Experiments on Maxwell's Theory 131

6. HERTZ'S EXPERIMENTS ON ELECTROMAGNETIC
WAVES. 135

6.1 Hertz's Initial Helmholtian Approach to the
Experiments on the "Exceeding Mobility of Electricity" 135
6.2 From Conduction of Currents in Wires to
Propagation of Changes of Potential 140
6.3 A Clue for the Academy Prize: Propagation of Very
Rapid Electric Oscillations in Short Open Linear Wires. 141
6.4 A Phenomenological Theory of the "Kreis" Detector 145
6.5 Hertz's New Conception in 1888: Waves of
Electric Polarization in Ether. 149
6.6 A Reinterpretation of the Experimental Results
in the Light of Maxwell's Theory 152
6.7 A Radical Change: from Charges and Currents in
Wires to waves and Fields in Space. 155
6.8 An Analysis of Hertz's Logic of Discovery. 159

7. HERTZ'S 1884 THEORETICAL DISCOVERY OF
ELECTROMAGNETIC WAVES. 167

7.1 Hertz's 1884 Principles of Uniqueness and
Independent Existence of Electric and Magnetic Forces
and his 1888 Discovery of Electromagnetic Waves 167
7.2 Why did Hertz not mention his 1884 Paper in 1888? 180

8. A FOUNDATION FOR THEORETICAL PHYSICS
IN HERTZ'S INTRODUCTION TO *DIE PRINZIPIEN
DER MECHANIK* 187

8.1 Hertz's *Bild*-Conception 187
8.2 Hertz's Criticism of Energetism 190
8.3 A Synthetic Representation of Mechanics. 193
8.4 A Comparison between Helmholtz's and Hertz's
Philosophies 195

9. ON BOLTZMANN'S MECHANICS AND HIS *BILD*-CONCEPTION OF PHYSICAL THEORY 201

9.1 Boltzmann's *Bild*-Conception and the Plurality of
Theories in the 1890's 201
9.2 Boltzmann's Mechanics and His Opposition
to Hertz's *Zulässigkeit* 206
9.3 The Historical Role of Boltzmann's Mechanics
and Gas Theory 211

PART THREE: FROM RELATIVITY TO QUANTUM THEORY

Foreword to Part Three 219

10. EINSTEIN'S CORRESPONDENCE CRITERIUM AND THE CONSTRUCTION OF GENERAL RELATIVITY 223

10.1 Correspondence as a Warranty of Continuity
with Tradition 223
10.2 What does Correspondence really mean ? 224
10.3 The Heuristic Role of CCr in Einstein's 1912
Construction of GR and the Relinquishing of the
Absolute Equivalence Principle 227
10.4. CCr Required a Sacrifice in 1913: Einstein's
Provisional Relinquishing of General Covariance 230
10.5. A Correspondence Criterion for the Operational
Definition of Space-Time Intervals 233
10.6 Stratification of Theories, Physical Reality and
Completeness 235

11. EINSTEIN'S LIFE-LONG DOUBTS ON THE PHYSICAL FOUNDATIONS OF THE GENERAL RELATIVITY AND UNIFIED FIELD THEORIES 239

11.1 A Foundational Problem in Einstein's Relativity........ 239
11.2 Einstein's Problem with the Stipulation of Meaning
for the Riemanian Space-Time Continuum 240
11.3 "Correspondence" as a Logical Asymmetric Method
for Correlating Concepts and Perceptions. Einstein's
Desire for a Purer Method............................ 242
11.4 Over-Determined Theories Avoid the SM Transgression.. 245
11.5 Einstein's Final View: the incompleteness of General
Relativity Seen as Lack of a Satisfactory Criterion for the
Postulation of Meaning............................... 247
11.6 Conclusions 250

12. CORRESPONDENCE AND COMPLEMENTARITY IN NIELS BOHR'S PAPERS 1925-1927 253

12.1 Karl Popper's Criticism of Bohr's Complementarity..... 253
12.2 Bohr's Correspondence Principle in 1925............. 255
12.3 From the 1925 "Restriction" to the 1927
"Limitation": Complementarity.......................... 260
12.4 Bohr's Deduction of IR 264
12.5 Bohr's IR in the Context of the "COMO" Congress..... 267
12.6 Bohr's Irreducible Disparity between Quantum and
Classical Theories..................................... 270

13. FROM THE 1926 WAVE MECHANICS TO A SECOND-QUANTISATION THEORY: SCHRÖDINGER'S NEW INTERPRETATION OF WAVE MECHANICS AND MICROPHYSICS IN THE 1950'S 273

13.1 Difficulties with the 1926 Wave Mechanics 273
13.2 Schrödinger's 1950's Remedy for the
Multidimensionality of the Psi Wave: Second
Quantisation 275
13.3 Schrödinger's Original Conception of the
New Statistics 280
13.4 Schrödinger's Criticism of Atomistic Ontology

in 1950 282
13.5 A Critique of the Copenhagen Philosophy............. 284
13.6 Continuity and Causality not at the Same
Foundational Level: a Two-Level Theory 288
13.7 Only the Ondulatory Theory is a "Complete" Theory 292
13.8 A Purely Theoretical Language Needs a Clear
and Precise Model...................................... 295
13.9 Schrödinger's New Ideas in the 1950's: a Programme
for New Physics 296

*Appendix 1. New Statistics and the Demand for a New
Objectification: Schrödinger's Simile Concerning Three
School-boys and their Three Awards* 299

*Appendix 2. Schrödinger's "Fuzzy" Cat is not Representable
by a Cloud but by a Factorisable Wave Functional* 301

14. CONCLUSIONS...................................... 303

Notes... 311

Bibliography: Primary Sources 351

Bibliography: Secondary Sources 363

Index... 379

INTRODUCTION

This book presents a perspective on the history of theoretical physics over the past two hundreds years. It comprises essays on the history of pre-Maxwellian electrodynamics, of Maxwell's and Hertz's field theories, and of the present century's relativity and quantum physics. A common thread across the essays is the search for and the exploration of themes that influenced significant conceptual changes in the great movement of ideas and experiments which heralded the emergence of theoretical physics (hereafter: TP).

The fundamental change involved the recognition of the scientific validity of theoretical physics. In the second half of the nineteenth century, it was not easy for many physicists to understand the nature and scope of theoretical physics and of its adept, the theoretical physicist. A physicist like Ludwig Boltzmann, one of the eminent contributors to the new discipline, confessed in 1895 that, "even the formulation of this concept [of a theoretical physicist] is not entirely without difficulty".[1] Although science had always been divided into theory and experiment, it was only in physics that theoretical work developed into a major research and teaching specialty in its own right.[2]

It is true that theoretical physics was mainly a creation of turn-of-the century German physics, where it received full institutional recognition, but it is also undeniable that outstanding physicists in other European countries, namely, Ampère, Fourier, and Maxwell, also had an important part in its creation. Moreover, since the beginning of nineteenth century, such French, English and Irish mathematical-physicists[3] as Poisson, Cauchy, Green, Stokes and Hamilton, by contributing analytical tools for the new discipline, paved the way to its mathematisation. At the end of century, Henri Poincarè in France paralleled Einstein's work in introducing new mathematical tools into the body of classical physics. A fruitful school of mathematical-physicists also flourished in Italy around that time. Therefore, with some notable exceptions such as that of the outstanding American scientist Williard Gibbs, one can speak of theoretical physics in the last century as a predominantly European discipline, though its methods and conceptions are now spread all over Western and Eastern countries.

The history of the origin and achievements of TP presents such a vast and variegated panorama that a single work can scarcely presume to cover the whole field.[4] My historiography is confined to the level of historical analysis that Gerald Holton, in his by-now classical book, *Thematic Origins of Scientific Thought: Kepler to Einstein*, labelled as the Z axis or "thematic dimension" of the historical enquiry.[5] I have attempted to explore the "thematic ideas" inherent in the foundations of TP by highlighting the contributions of some great physicists who preceded and followed the eventful changes from nineteenth-century mechanics to electrodynamics and the present century's TP.

I have focused my attention on three themes which run through the three parts of the book. One is the role of mathematisation in the transition from nineneth-century mechanics to the sciences of electrodynamics and electromagnetism. Another is a new perspective on the theory-experiment relationship during this transition. Finally, in my third theme I investigated the contribution of the phisicist's philosophy in shaping their theoretical approaches.

Mathematisation of electrodynamics is an appropriate topic to begin my analysis. Mathematics made a fundamental contribution to the process which led to the affirmation of the new discipline, a contribution that was recently labelled[6] "the torch of mathematics". I argue that mathematisation in physics was far from being unidirectional (as if it were inspired by a unique perspective) in its methods and aims. My papers show that it took various forms, ranging from Ampère's and Weber's algebraisation of laws of physics to Maxwell's emphasis on mathematical analogies and dimensional analysis, not to mention Einstein's non-Euclidean approach to general relativity and the use of operators in the formulation of quantum theory.

In his field theory of electromagnetism, Maxwell conceived of an analogical correlation of mathematics with physics which I argue changed the theory's status from a mere description to a model of physical reality. In this sense, I consider Maxwell's theory to be the forerunner of the theoretical physics that developed in the final quarter of the nineteenth century.

Mathematisation of physics was also responsible for a new view on the theory-experiment relationship.[7] Maxwell and Hertz, by introducing into electromagnetism the mathematically and physically powerful partial differential equations of elasticity and hydrodynamics, were confronted with the fact that their theories were becoming increasingly remote from an empirical basis. I argue that Maxwell's dynamical theory and Hertz's *Bild* - conception of theories can be interpreted as new strategies devised to solve this problem. The radicalism of these strategies can be deduced from Maxwell's, Hertz's and Boltzmann convictions that experiment was not a crucial test for theory's validation.

The issue of a reciprocal influence between physics and philosophy has been recently examined by philosophers, scientists and historians although with different aims and perspectives. The word "influence" has a variety of meanings, involving social, metaphysical and methodological factors which differ across disciplines.[8] In my study I have tried to demonstrate that, throughout the second half of the nineteenth century, physicists such as Helmholtz, Maxwell, Hertz, Mach, and Boltzmann were all aware of some philosophical problems implicit in their choice of theories. In the first decades of this century, Einstein, Planck, Bohr, Heisenberg, Schrödinger, just to mention a few in an outstanding list, were also aware of similar problems. Ernst Cassirer has convincingly illustrated [9] one aspect of these problems. The historical development of TP brought to light a contrast between two tendencies which were active within physics at least since the time of Galileo and Newton. On the one hand, there was a tendency to generalise theory, making it maximally comprehensive to encompass all phenomena of nature - thus reaching a unique theory of the physical world; and, on the other hand, an inclination to emphasise the empirical level and to develop phenomenologically limited theories.

In their desire for unity, physicists of the first tendency encountered problems in the methods and scope of their research. Their speculations on these matters naturally and consequently poured over into philosophy. Newton's *Regulae Philosophandi* in his *Principia* was an early example thereof; a recent example is Einstein's struggle to construct general relativity.[10]

The opposite tendency was expressed by physicists who thought empirical science had to be restricted to facts and that this could most surely be done by avoiding all subtle epistemological reflections and speculations. The impossibility of keeping faith with this program of *naive realism* [11] is demonstrated, among other things, by the physicists' epistemological contributions quoted in this book.

One can thus argue that the physicists' general ideas and broad conceptions were actually reflected in their theoretical and experimental achievements. As Helmholtz once said, there was a time when the fundamental problem at the beginning of all science was the problem of epistemology: "What is true in our intuition and thought?". He believed that philosophy and science confront this problem from two opposite sides, so that this task is common to both.[12] In my essay I show that, according to his own testimony, Helmholtz regarded his work with the non-Euclidean geometries as an improvement on Kant's doctrine of the a-priori (not a refutation, as is often mistakenly thought).

Many chapters of this book illustrate the thesis that foundational problems in physics have been tackled through the invention of new instruments and/or new conceptual tools, be it new forms of mathematics (Newton's calculus is the main example thereof), or new methods of connecting complex-number analysis and imaginary numbers with physical quantities (i.e., Hamilton' and Maxwell's quaternion calculus, Heisenberg's matrix-calculus, etc.). Yet these tools became most fruitful when their usage was guided by new ideas, i.e., by innovative scientific programs and broad conceptions.

I assembled the essays in three Parts following their historical sequence. Part One concerns the impact on the mathematisation of electrodynamics exerted by Ampère's, Gauss's and Weber's theories. Ampère was the first to trust the mathematical form of his electrodynamic laws more than their physical content. Through their metrological program of absolute systems of units, Gauss and Weber took the next step of writing electrodynamic laws in the form of algebraic equations, thus more tightly connecting their mathematical form with the physical content. Maxwell extended the process by introducing into electromagnetism the mathematically and physically powerful partial differential equations of elasticity and hydrodynamics.

In my critical evaluation of Hertz's researches, the principal subject of Part Two, I have mainly stressed the importance of his 1884 theoretical paper, which is often considered by Hertz's historians (not, however, by Planck) as parenthetical to his contributions to electrodynamics. Hertz's 1884 contribution was a rather formal and mathematical theory, lacking its own corresponding physical conception of propagation in the ether. The latter conception as it already existed appeared convincing to Hertz when he started the experimental research that was meant to confirm Helmholtz's theory of the polarization (*à la* Poisson) of material dielectrics and ether. However, Hertz soon discovered that Maxwell's waves were not waves *à la* Poisson. This discovery culminated in one of the most masterful achievements in the history of modern physics: Hertz's renowned experiments on the finitely propagated electromagnetic waves in air.

In Part Three, I analyse Einstein's ideas on the problem of the new relationship in his theories of relativity between concepts and empirical data. I touch on a specific feature of Einstein's epistemology, taking my cue from his ways of generalising the special theory of relativity and Newtonian gravitational theory into a general theory of relativiy. I argue that the so-called correspondence requirement represented for Einstein the adoption of a technical tool, supported by a general postulate of continuity in the methods of physical research. In this role, correspondence guided Einstein as he rapidly developed such a highly mathematised discipline as general relativity. I explore Einstein's convictions and perplexities when faced with the doubts posed by the effective significance of correspondence. Bohr elevated correspondence to the role of a principle, and I look at the relation between Bohr's actual usage of this principle and his Copenhagen philosophy. In my last Chapter on Schrödinger's contributions in the 1950's, though I agree that Schrödinger's later works lacked a fully fledged theory founded on his epistemological conceptions, my explanation of this absence is rather different from that presented in most parts of Schrödinger's literature.[13]

Although each essay in this book is self-contained, a sequential reading of the whole work allows a more comprehensive view of the historical development of theoretical physics in the last two centuries.

Summing up, this work presents a break through a great movement of ideas in the panorama of nineteenth century physics, which, in my view, conditioned the rise and growth of theoretical physics. What was at stake, at bottom, concerned fundamental issues of western culture, such as the idea of an empirical science and the role of experience in the acquisition of knowledge.

I believe that the reader of this book can be confronted with an alternative between two contrasting views on the influence of the physicists' philosophy on their theories and experiments. One is the view that the two processes of discovering new facts, and of changing ways of "thinking physics" always intersected each other. This implies that the philosophical discussion has been an essential, not marginal, component of the advance of physics.

Evidently, this view is contradicted by those who consider the above issues marginal to the great achievements of theoretical and experimental physics in the same period of time, an epiphenomenon with respect to "real" science. Consequently, they believe that historical changes in the way of "thinking physics" are irrelevant for physics itself, because scientific method, alike with nature, has been in essence always the same. New theories and discoveries of new facts about nature are the norm in history - this is conceded - but the way of "thinking physics" has remained the same, since Galileo, at least.

I hope that whether the reader chooses between these views, he will continue to manifest his benevolence to the author.

Salvo D'Agostino
Roma, August 1999

PART ONE

From Mechanics to Electrodynamics

FOREWORD TO PART ONE

Part One of this book deals with the history of electrodynamics and electromagnetism from Ampère to Maxwell and Hertz.

Ampère founded his *Electrodynamique* on a force law between the so-called elements of currents involving both galvanic and magnetic forces. By connecting galvanism and magnetism into a unique term, this inverse-square-of-distance law represented the first successful attempt at unifying formally, i.e., mathematically, electric and magnetic forces. He rediscovered the law through a new methodological approach: the search for symmetries in the equilibrium of conducting wires, a method which, being purely geometrical, was for him free from historically conditioned physical hypotheses. Consequently, he claimed perennial validity for both his law and his method, stating that his method was the continuation of the Newtonian method, summarised in the words "hypotheses non fingo".

The next step in mathematisation was achieved when Gauss and Weber developed a keen interest in this new science. Through their new approach, founded on the metrology of the absolute systems of electric and magnetic units, electrodynamics received in their hands, a complete (for that time) mathematical vest.

I find it remarkable that, due to the above systematic organisation of units in Weber's theory, laws could be written in the form of analytical equations, including physically significant proportionality constants. One of these constants conquered the role of a second universal constant. Added to the Newtonian gravitational constant, it conferred a remarkable conceptual role on the Weberian technical approach to metrology.

In another essay, I discuss the story of Maxwell's life-long efforts directed at proving that Weber's constant really measured Maxwell's propagation velocity of electromagnetic waves and of light, not, as Weber had claimed, the velocity of the motion of electrical particles. In this proof, Maxwell brought to his electromagnetic theory of light concepts and data that were developed in the 1850's by Weber. Maxwell's conclusion was that Weber's conversion factor for electrical units, which in Weber's theory corresponded to the relative velocity of motion of electric particles, represented the velocity of electromagnetic waves and of light in an electric ether. A momentous conclusion because it transformed the relative velocity of motion of particles into the absolute velocity of electromagnetic waves!

The conceptual difficulties presented by this transformation are well documented by Maxwell in papers he wrote over many years. In 1861-62, the connection between Weber's factor and the velocity of the electromagnetic waves turned out to be dependent on an "ad hoc" hypothesis, the adoption of an appropriate hydrodynamic elastic model for ether. The electromagnetic wave velocity was also derived in analogy with the velocity of an elastic wave. In 1863, Maxwell's recourse to a theory of two absolute system of units provided a better way of relating Weber's factors with the ethereal constants. During the same period he collaborated in the activities undertaken by the Committee on Electrical Standards. In 1864 the novelty of the above transformation resided in deriving the electromagnetic wave's velocity from a D'Alembert-type equation, and his method implied a certain degree of immunity from the elastic argument mentioned above. By 1868 an important stage in his struggling with the transformation was reached: the identification of Weber's factor with the constant ratio between electrostatic and electromagnetic forces.

In *A Treatise on Electricity and Magnetism* (1873), Maxwell performed a grandiose operation: the construction of a complete metrological theory of two absolute systems of units for electric and magnetic quantities, as well as the development of a consistent theory of dimensions for these quantities. The operation gave Weber's factor a predominant function in the metrological theory. In fact, as a consequence of this operation, Weber's factor appeared consistently in every dimensional and numerical relation between different units for the same quantity in the two systems. The equality between Weber's factor and the light velocity was then demonstrated in simple, formal mathematical proofs.

Maxwell's strenuous efforts to insert Weber's constant in a completely modified theoretical context were supported by his new approach to physical equations, i.e., by his idea that these equations are bearers of "dimensions". This approach was consistent with Maxwell's ideas of a mathematics "embodied" in physics and represented a remarkable advance towards theoretical physics.

Chapter Three as well as my short essay on the history of units and dimensions can be considered as original illustrations of the influence on the development od theoretical physivs that resulted from metrological innovations, which have often been considered merely technical procedures.

In the following pages, I explore the problems posed by the growth of theoretical physics in the second half of the last century which were debated by the great masters of physics of that period,

Maxwell, Helmholtz, Hertz and Mach. Some of these problems concerned the new form of relating observations to the physical concepts initiated by Maxwell's analogical view of theory: the lack of uniqueness in the system of concepts, or theoretical pluralism. This position posed a threat to the old conception of a descriptive theory, which Maxwell provisionally exorcised. Helmholtz' parallelism of laws and Mach's phenomenism were also a way of dispelling the threat of pluralism. However, the issue of theoretical pluralism was revived in Hertz's philosophy, while a method for rational criteria of choice among empirically equivalent theories was one of Einstein's concerns.

CHAPTER 1

A CONSIDERATION ON THE CHANGING ROLE OF MATHEMATICS IN AMPÈRE'S AND WEBER'S ELECTRODYNAMICS

1.1 Ampère's Electrodynamics

In 1820 Andre Marie Ampère (1775 Lyon - 1836 Marseilles) coined the term électrodynamique to indicate that the new science of electric currents and magnets was part of the Newtonian program of a general science of forces and motions. Adopting the spirit of Ampère's works, Gauss and Weber translated the French into *Elektrodynamik* and Helmholtz and Hertz also used this German term in their reinterpretation of Maxwell's ideas. Conversely, the term "*electromagnetism*" appears to have originated with Oersted and, extensively used in Faraday's and Maxwell's work, remained the standard word in the Anglo-Saxon tradition. Following the German tradition, Lorentz and Einstein used *elektrodynamik* while most others theoretical physicists of the twentieth century preferred the English translation of the term, "electrodynamics".

Ampère presented a systematic exposition of his electrodynamic theory in his "Théorie des phénomènes electrodynamiques uniquement déduits de l'experience" (Paris,1826). In his research, Ampère investigated the equilibrium conditions in the mobile part of an electric circuit (le conducteur mobile) subjected to the combined actions of other parts of the same circuit or of other circuits, when the form and the positions of the circuits were varied. The above conditions allowed him to deduce a law of the electric force acting on the elementary portions of any electric circuit, which he called the elementary law of electrodynamic force. Since he balanced the forces exerted on a part of a circuit with the forces exerted by the whole circuit or by other circuits, his instruments were traditionally named electric balances.

Ampère believed in the possibility of obtaining a general law of electrodynamics by adopting a special method, his second method. He introduced it in the following passage:

Mais il existe une autre manière d'atteindre plus directement le même but (i.e. the electrodynamic action law)...elle consiste à constater par l'expérience, qu'un conducteur mobile reste exactement en équilibre entre des force égales, ou des moments de rotation égaux, ces forces et ces moments étant produits par des portions de conducteurs fixes dont les formes ou les grandeurs peuvent varier d'une manière quelconque, sous des conditions que l'expérience détermine, sans que l'équilibre soit troublé, et d'en conclure directement par le calcul quelle doit être la valeur de l'action mutuelle de deux portions infiniment petites, pour que l'equilibre soit en effet indépendant de tout les changements de forme ou de grandeur compatibles avec ces conditions.[1]

To unambiguously determine his law, he conducted four experiments on closed circuits. Although his experiments dealt with closed macroscopic circuits carrying electric current, his elementary law of force concerned forces between *infinitesimal* parts of the circuits, his *éléments de courants*. When integrated along the real circuits according to the rules of analysis, this elementary law would yield the total force acting between the complete circuits. Ampère's analytical approach to the problem of interaction between currents was retracing the Newtonian approach to the law of universal gravitation.

Let us briefly review the four experiments that contributed to

Ampère's fame in the history of nineteenth- century science.

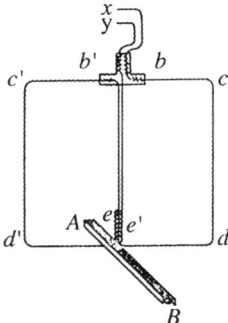

Fig. 1. First apparatus of Ampère's derivation

The first apparatus (Figure 1) consists of an "astatic circuit" (i.e. an electric circuit with two loops where the direction of current flow was inverted to balance the effect of the magnetic field of the earth). It remains in equilibrium under the action of a current flowing in a segment of a rectilinear conductor having its centre on the axis of rotation of the circuit. The equilibrium persists when both currents invert their directions, thus proving that the force is an even

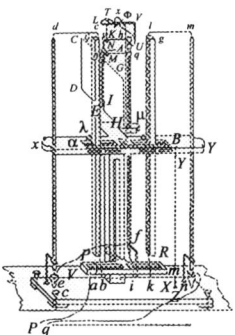

Fig. 2. Second apparatus of Ampère's derivation

function of the currents, i.e., it is invariant under the inversion of both currents.

The second apparatus (Figure 2) shows that the "astatic" circuit remains in equilibrium when a wire twisted into small sinuosities takes the place of the straight wire. Ampère interpreted the experiment by posing the thesis that the force exerted by an element of current has the transformation properties of a vector. The results of the first two experiments allowed Ampère to write the elementary law in the form:[2]

$$d\mathbf{B} = a\,\mathbf{r} \tag{1}$$

$$a = ii'\{\mathbf{ds}\,\mathbf{ds'}\phi(\mathbf{r}) - (\mathbf{ds.r})(\mathbf{ds.r})\psi(\mathbf{r})\}$$

where $i\ i'$, signifies current intensity; and **ds ds'** signifies *vectors representing elements of length of the circuits*. For brevity's sake, I give a modern rendering of Ampère's actual notation, transcribing his formulae in modern vectorial notations (bold face letters represent vectors).

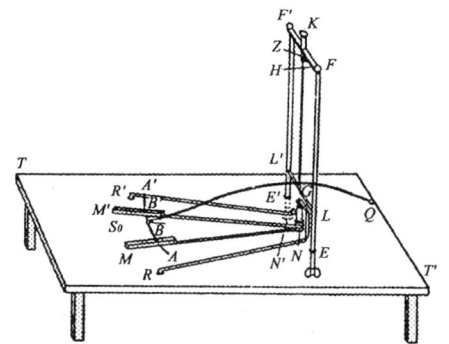

Fig. 3. Third apparatus of Ampère's derivation

The third apparatus (Figure 3) is more complicated and Ampère presented it in different arrangements. One of last arrangements, referred to in an 1827 memoir, consists of a sophisticated artifact in which electric currents are forced to follow complicated paths. A section of a circuit free to rotate around an axis remains in equilibrium when brought near a second closed circuit. Ampère interpreted the result as the proof that the resultant force, exerted by a closed circuit on a mobile element of another, is always at right angles to the latter, i. e., as the proof of the law that no tangential forces are exerted between the interacting electric wires. This law entails a symmetry: the equilibrium of a circular circuit, free to rotate around a fixed axis, is not disturbed by the action of a second circuit of whatever form. Ampère's interpretation of the equilibrium conditions in this third experiment represented an important achievement for the theory of electrodynamic action; among others, Maxwell derived many consequences from this type of equilibrium.

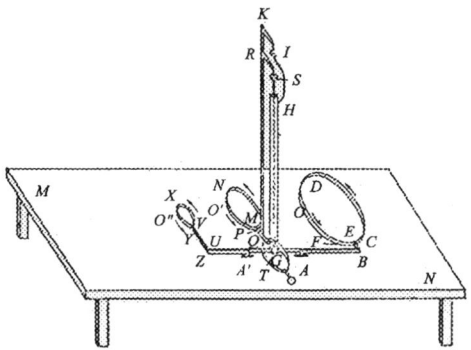

Fig. 4. Fourth apparatus of Ampère's derivation

In the fourth and final experiment (Figure 4) we have three flat circuits all elliptical in shape; two, O and O", are fixed, while the third, O', lying between the two fixed ones, is movable. O has dimensions n times greater and O" n times smaller than the movable O'. They are similarly positioned in the sense that the distance between O and O' is n times greater than the distance between O' and O". The equilibrium of the central circuit under the combined action of the other two proved, according to Ampère, that the force between two elements was unaffected when all the linear dimensions of the system of circuits are proportionally increased, the current strength remaining constant. Another argument proved to him that this rule can be generalized to circuits of whatever shape. That the equilibrium of currents remains invariant under similar transformations (a scaling law) proves that their interactions follow an inverse square law.

In consequence, the two functions are:

$$f(\mathbf{r}) = A/r^3 \;\; ; \;\; y(\mathbf{r}) = B/r^5$$

The rotation invariance allows us to find:

$$B = -3/2 \, A$$

In short, Ampère's second method consisted of modifying the geometrical properties of the systems of currents in such a way as to preserve the equilibrium conditions of a portion of the circuit. Specific symmetry properties of the elementary law were deduced from the invariance of the equilibrium of finite parts of the circuits when their shape is varied. In so doing, he implicitly assumed that the elementary law of force possessed the same degree of freedom as the corresponding equilibrium conditions, i.e., equilibrium under reversal of currents, vectorial type of equilibrium, rotation equilibrium and equilibrium in similar transformation. These four types of equilibrium and the corresponding symmetries were considered by Ampère as necessary and sufficient conditions for the unambiguous determination of the elementary law of force.

1.2. Ampère's Theory of Equivalence and its Foundation on the Symmetries of Magnetic Forces

Given Ampère's interest in the unification of forces, it is understandable that he extended his symmetry arguments from electric to magnetic phenomena in attacking the problem of the nature of magnetic forces.

The result was his well-known electrical theory of magnetism, the so-called theory of equivalence. It amounted to an extension of Ampère's arguments on symmetry that he used to formulatehis electrodynamic law. In sum, Ampère based his reduction of magnets to currents,[3] the celebrated electric theory of magnetism, on the same symmetry argument about symmetry employed to derive his electrodynamic law. Suffice it here to quote a passage from the 1822 memoir "Sur la détermination de la formule qui représente l'action mutuelle de deux portions infiniment petites de conducteurs voltaiques". Starting from the theorem of tangential actions discussed above, Ampère extended his arguments from circuits to magnets, affirming that:

> Il en devait être de même d'un assemblage quelconque de circuits fermés, et, par conséquent, d'un aimant, lorsqu'on le considère comme tel, conformément au mon opinion sur la cause des phénomènes magnètiques, et c'est, en effet, ce qui resulte de plusieurs experiences dues à divers physiciens.[4]

Magnets are currents because they have the same symmetrical properties as currents have. The same formal arguments serve him to disprove the opposite theory of the reduction of electricity to magnetism, a theory which had been advanced by circles close to Örsted and to the German scientists. This theory explained Örsted's experiments by the hypotheses that small transversal magnets are located around the conducting wire, as supposedly proved by the motion of the compass near the wire. For this reason this theory was labelled "la théorie de l'aimantation transversale".[5]

Ampère's line of reasoning conceded that, in principle, a theory of transversal magnetism (all the currents are magnets) was admissible. However, this theory should be submitted to an experiment showing that all the properties of symmetrical interactions between currents were also properties of interactions between magnets (and not the reverse).

The underlying rationale of this argument can be stated as follows: if, according to the theory of transversal magnetism, currents are magnets, the possible interactions among magnets must form a larger set than the interactions among currents; in fact, some types of interactions among free magnets would be impossible in currents. In short, if the theory of transversal magnetism were to be proved, the set of interaction among magnets should have been larger than the set of interactions between currents (Figure 5).

If vice-versa, as Ampère asserted, magnets are currents the reverse

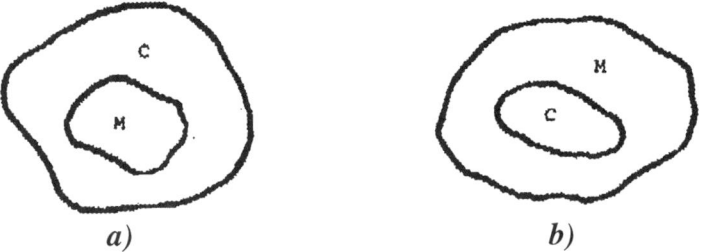

Fig. 5. Logical application of the two theories:
  a) Magnets "are" currents (Ampére's theory of equivalence).
  b) Currents are magnets (Theory of transversal magnetism).

case should occur, i.e., the set of possible interactions among currents should be more extensive, and types of interactions must exist among currents - or among currents and magnets - which do not occur among magnets alone.

According to Ampère, the experiment proved this second case, because a continuous rotation motion is feasible only with currents and magnets, while, on the contrary, nobody has ever seen a couple of magnets rotating one around the other.⁶ The supporting evidence was provided in an experiment (Fig. 6) performed by

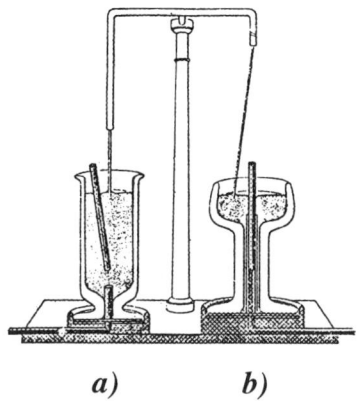

Fig. 6. Faraday experiments:
a) Rotating magnet;
b) Rotating current.

Michael Faraday. On the left side, it showed that a magnet was rotating in the field of force of a current, and, on the right side, that a conducting wire was rotating in the field of a magnet. But no rotation of a magnet around another magnet was observed.

1.3 Ampère's Program : Only Mathematical Laws are Perennially Valid Because they are Independent from Physical Hypotheses

Ampère stated why he prized his method so highly: he had deduced his law from the equilibrium conditions and geometric symmetries of his circuits. This deduction was trusted to perennial mathematical considerations and not to passing physical hypotheses:

Quelle que soit la cause physique à laquelle on veuille rapporter les phénomènes produits par cette action (électro-dynamique), *la formule que j'ai obtenue restera toujours l'expression des faits.* Si l'on parvient à la déduire d'une des considerations par lesquelles on a expliqué tant d'autres phénomènes, telles que les attractions en raison inverse du carrè de la distance, celles qui deviennent insensibles à toute distance apréciable des particules entre lesquelles elles s'exercent, les vibrations d'un fluide répandu dans l'espace, etc., on fera un pas de plus dans cette partie de la physique; mais cette recherche, dont je me ne suis point encore occupé, quoique j'en reconnaisse toute l'importance, ne changera rien aux résultats de mon travail, puisque pour s'accorder avec les faits, il faudra toujours que l'hypothèse adoptée s'accorde avec la formule qui les représente si complètement.[7] [Italics added]

It can be argued that Ampère considered physical causes as historical accidents which complemented but could never challenge the validity of a purely formal mathematical law. In his criticism of physical hypotheses concerning the causes of phenomena, Ampère's targets were presumably Poisson's inverse square law and Örsted's explanation of the magnetic effect of currents. Let us notice that, in his opinion, the form of his law was deduced from factual observations. On the contrary, any physical conception forwarded as an explanation of the phenomenon, is, for him, an addition to the pure facts of experience. Only the "form" of the law is uniquely supported by "experience", i.e., by his experiments with the electric balances.

Leaving for my final chapter a critical appraisal of Ampère's ideas, let me briefly note that Ampère's mathematical physics was in the tradition of D'Alembert and Lagrange's Mechanics and that it was soon enhanced by Fourier's great approach to the theory of heat propagation. The methods of the French mathematical physicists were somehow defeated by Faraday's methods and by Maxwell's analogical approach to theory.

It seems that Ampère was inclined to consider magnetic forces as a modification of electric forces due to the motion of electric charges.[8] Incidentally, this was a physical hypothesis, methodologically inconsistent with the above self-prescribed rule. However, perhaps for this reason, he did not work out this theory. After ten years, Weber in Germany was inspired by Ampère's ideas.

1.4. Gauss's and Weber's Metrology. Weber's Discovery of a Constant Velocity of Motion of Electric "Masses"

Carl Friedrich Gauss (1777-1855) brought about a "revolution" in his research in earth magnetism[9] because in his own words [10] he was "*a practical astronomer i.e., one of those mathematicians who are familiar with the finest means of observation*". It was in fact this combination of high mathematical expertise and great experimental ingenuity that characterised Gauss's research method.

Throughout 1832 Gauss worked to develop and test a method for measuring the quantity and direction of the earth's magnetic intensity, independently of the characteristics of the measuring compass. In fact, the methods used until then were largely unreliable mainly because the measures were dependent on the particular magnetic moment of the compass employed and were variable in time due to variations in this moment. The results were therefore unreliable for any investigation of the earth's magnetic variation at a given location over a long stretch of time. Furthermore, these methods afforded only relative measures that were not comparable with others from different locations and different instruments.

Gauss's method, the so-called method of "principal positions", is still described in many textbooks of elementary physics, so that only a short description is given here.

The gist of the method was the usage of an auxiliary instrument in addition to the newly constructed high-precision bifilar magnetometer. Gauss used a sensitive compass suspended from a torsion-free long silk wire (magnetometer) located in a special position with respect to the bifilar magnetometer. Through the combination of the data of the two instruments, Gauss succeeded in eliminating the magnetic moment of the bifilar magnetometer, thus finding direction and horizontal intensity of the earth's magnetic force.[11]

Understandably, the comparison of these measurments with others from different locations was improved when these data were related to the mechanical fundamental units of mass, space and time, for these were rather easily transferable to other locations.

Due to both features (independence of instrument and independence of location), Gauss called the units "absolute". Both the magnetometer and the bifilar magnetometer were constructed by Gauss with new criteria, which allowed a sensitivity and precision never achieved before in geodetic instruments and only comparable to the precision achieved in astronomic instrumentation.[12]

Wilhelm Eduard Weber (1804-1891) was, for a major part of his life, a collaborator and a friend of Gauss at the University of Göttingen. The path to Weber's electrical researches lay through Gauss's magnetism at Göttingen. In fact, Gauss's and Weber's magnetic interests soon extended to the exploitation of the magnetic techniques in the new field opened by Faraday's recently discovered electromagnetic induction. For this purpose Weber constructed his induction inclinator (1837) and rotation-inductor (1838).

In the following years Weber extended Gaussian absolute measurements to the entire field of electricity and magnetism. One of Weber's first essays in this direction involved comparing the magnetic field produced by a current with the earth's absolutely measured magnetic intensity in order to obtain the definition of an absolute magnetic unit for a galvanic current.

From the differential form of the Biot-Savart law, it followed that the magnetic force at the centre of a circular loop of radius R conducting a current i, would be $2\tau i/R$. When in the vertical meridian plane of the earth's magnetic force H, this loop exerted a deflecting force on a small pivoted compassneedle, located in the centre of the circle (equilibrium of the needle occurred at the angle ϕ, where Tan $\phi = 2\pi i/HR$). The absolute magnetic unit of current was equal to the current that, in the situation above, deviated the needle by an angle ϕ, such that Tan $\phi/2\pi = 1$ when R and H are themselves equal to one in absolute units. In order to measure *Tan* ϕ, Gauss and Weber constructed in 1840 a very accurate tangent galvanometer,[13] which was soon transformed into a dynamometer when the central compass was replaced by a bifilar suspended multi turn currentcoil.

In 1843 these activities were interrupted when, for political reasons, Weber was compelled to leave Göttingen University.[14] He soon joined the faculty at Leipzig University where Gustav Theodor Fechner (1806-1887) was actively researching electric currents and physiology. Weber accepted and included in his theory. Fechner's special view of an electric current consisting in one wire of a double motion in opposite directions of positively (+e) and negatively (-e) charged particles (Weber: 'electric masses'). In the other wire, these particles (indicated as +e' and -e') moved in directions which were opposite to each other and also opposite to the direction of like-sign particles in the first wire.[15]

In 1843, Weber had become particularly concerned with Ampère's electrodynamics. When he returned to Göttingen in 1849 he had already contributed to electrodynamics important results which were published in 1846. This research culminated in the discovery of a fundamental law of electrodynamic action, which he presented in his influencial 1846 paper " Electrodynamic Measures on a General Fundamental Law of the Electric Action".[16]

Weber was convinced that Ampère's theory had a future if somebody took on the task of refining his measurements, completing the theory, and extending the observations.[17] For Weber, Ampère could not support his claim that his law was "derived only from experiments", because, in the connections between the movable conductor and the battery, friction perturbed the electrodynamic force that was to be measured.

In order to reduce friction, Weber constructed the "electrodynamometer" (or just "dynamometer"), an instrument that he extensively used in his experiments.[18] The movable part, the so called "bifilar-roll", consisted in a wire coiled around a thin wooden frame, which was suspended by two fine metal wires conducting the current to the coil. The bifilar-roll was set in motion by the magnetic force exerted by a current circulating in a fixed coil of wire, the 'multiplicator'.

Through his bifilar electrodynamometer, Weber accurately measured the torque between the two current-carrying coils, thus testing the validity of Ampère's law of force between currents. He found that Ampère's law about current elements was correct.[19] The same instrument was used to prove that Ampère's law was valid also for "ordinary electricity", i. e., for a discharge current from a Leyden jar.[20]

A glance at Weber's work with the ballistic performance of his electro-dynamometer might be helpful at this point as an example of his method for relating theory to instrumental operations. When the same current flowed for a short period through both coils of his electro-dynamometer, Weber's instrument responded ballistically to the square of the current, i.e., to $\int i^2 dt$. On the other hand, the suspended magnet of the galvanometer responded ballistically (for example in the discharge of a Leyden jar) to $\int i dt$. A comparison of the two measurements [21] allowed Weber to estimate the duration of a current discharged by a Leyden jar through various lengths of a wet string, an operation which soon acquired significance for his later work in the measurement of his characteristic velocity. The quadratic response allowed also Weber to measure the rapid or high frequency currents detected in acoustic and physiological investigations.[22]

Another significant application to the measurement of variable current was the following: currents induced in the stationary coil of the electrodynamometer by the oscillations of the current conducting bifilar-roll produced a damping in its oscillations which was compared to the damping produced when " an equivalent magnet (i.e., equivalent from an electromagnetic point of view) " replaced the current carrying bifilar-roll. The equal amount of damping in the two experiments confirmed both Ampère's law of equivalence (between magnets and systems of closed, current bearing loops) and Faraday's induction law.[23]

Weber's starting point for his discovery of the fundamental law of electrodynamics was his conviction that electromagnetic induction, discovered by Faraday, was an indication that the electric fluids exert ponderomotive forces between wires (the electrodynamic actions studied by Ampère) precisely because of actions between the fluids themselves, which included as a particular case the Coulomb electrostatic forces. At best then, Ampère's law could only be a partial theory.[24]

It is no surprise then that Weber's next step was to search for a fundamental law which would encompass both Ampère's and Faraday's discoveries and take into account the special case of electrostatics.[25] The law was to be considered fundamental in the sense that it applied to the electric "masses" themselves rather than to their ponderable carriers, the conducting wires. These assumptions underlay Weber's 1846 paper in which he presented the fundamental law of electrodynamics (*elektrodynamische Fundamentalgesetz*).

Weber developed an extended theoretical and experimental study of electric charges, of currents and of their forces of interaction, culminating in his discovery of the fundamental law (henceforth: FL) of electric interactions. He constructed his FL from what he considered to be the virtually certain interactions between Ampèrian current elements. Although purely virtual (because elements of current are not physical objects in Weber's system) it was the successful experimental test of their integrated effects that convinced Weber of their "certainty", hence of their "factuality".[26]

In fact, he listed three "facts" (*Tatsachen*) that represented particular cases of Ampère's law, corresponding to particular positions of the current elements and of the conducting wires:

First Fact. Two parallel current elements whose directions lie on the straight line joining them, initially repel or attract as the electricities in them move in the same or in opposite directions.[27]

The other two "facts" concerned the actions between two parallel current elements, lying perpendicular to their joining line, and the current induced by a current element on an element of wire lying on the same line.

These three facts should be regarded as electric, i.e. one considers the exerted forces as reciprocal actions of electric masses.

Weber also assumed that the electrostatic force was weakened by a factor $a^2(\frac{dr}{dt})^2$ because of the relative motion of an electric 'mass' e in one wire and a 'mass' of equal sign e' in the other. The total force became:

$$ee'/r^2(1 - a^2 (\tfrac{dr}{dt})^2).$$

By 'fact' three and other experiments (in particular the electromagnetic induction), the forces between the electrical 'masses' should also depend on their reciprocal acceleration $\frac{d^2r}{dt^2}$, so that the complete law took the form:[28]

$$\frac{ee'}{r^2}(1 - a^2(\tfrac{dr}{dt})^2 + b\tfrac{d^2r}{dt^2}).$$

The first term ee'/r^2, i.e., the static force between e and e', was written without multiplicative constants and this implied, as Weber pointed out, the selection of appropriate units for the measures e, e', of the charges, units which he labelled mechanical units of charge. The dynamic part of the force is composed of the current-current force:

$$-\frac{ee'}{r^2} a^2 (\tfrac{dr}{dt})^2,$$

and of a force depending on acceleration:

$$- \frac{ee'}{r^2} b \frac{d^2r}{dt^2}.$$

In order to compare these two forces with special forms of Ampère's

law (i. e. forms of this law occurring in Weber's three facts), Weber used[29] expressions connecting $\frac{dr}{dt}$ and $\frac{d^2r}{dt^2}$ with Ampère's elements of current *ids, i'ds'*, and with the velocities, *U, U"* of the electric masses with respect to the wires (Weber: "absolute velocities").
The comparison allowed a reduction of constants: $b = 2ra^2$.[30] The FL was thereby presented in a neater form:[31]

$$\frac{+-ee'}{r^2}\{ 1 - a^2 (\tfrac{dr}{dt})^2 + 2a^2 \tfrac{d^2r}{dt^2} r \} \tag{1}$$

The plus sign referred to forces between pairs of particles, *e* and *e'* of like signs, the minus sign to pairs of different signs.

Weber's next step was to find a relation between the constant *a* in his FL and his definition of units of current intensity in various systems of units. He defined as the measure of current intensity the amount of like signs particles carried per second in one single flux in a single wire (let us recall that in Fechner and Weber's assumptions a current consisted in a double flux of charged masses in a wire); consequently, as one unit of current intensity he defined the current which carried unit charge per second in a single flux of electrical 'masses'. Consequently, Weber expressed the relation between the current intensity *i*, the linear charge density $e^* = e/ds$, and *U*, the 'absolute velocity' (i.e., velocity of electric masses with respect to the wire) as:

$$i = me^*U .$$

(In effect, Weber wrote[32]: $i = ae^*U$, but he distinguished his *a* here from his *a* in the former expression of FL. To avoid confusion, I prefer to use *m* in place of Weber's *a*). According to Weber, e^*U is the measure of current intensity in mechanical units; therefore, *m* represented a ratio dependent on the selection of particular units for i. When this definition of current intensity and the above transformation for passing from Ampère's ids to Weber's $e\,dr/dt$, are substituted in Ampère's law in order to obtain the FL, m appeared as the factor in a new form of the expression of the FL. In the case of repulsive forces, this expression[33] was:

$$\frac{ee'}{r^2}\{1-(m^2/16)(\tfrac{dr}{dt})^2+(m^2/8)(r\tfrac{d^2r}{dt^2})\}. \qquad (2)$$

By comparison with formula (1) above, it clearly appears that a in (1) depended on the ratio m, i.e., on a particular choice of units for the current, and that, therefore, it could not be considered the inverse of a constant physical velocity. However Weber was convinced that such a velocity existed and the search after it was his main task in the following years.

Since no constant velocity is included in Ampère's law, its relationship with Weber's FL presents some interest. Weber asserted that his FL represented "a generalisation of that previously erected by Ampère, which in effect represented the special case of four electrical particles simultaneously involved, when current elements are assumed constant and fixed".[34]

As is evident from these remarks, Weber did not deduce his FL from Ampère's law but he extrapolated it from Ampère's, limiting the indeterminacy of such procedure by testing his FL for particular cases, where it was amenable to "certain" experimental facts. Thus, his method consisted in a back and forth connection between electromagnetic experiments (through his instruments) and particular cases of Ampère's law. Given this situation, it is understandable that he wanted to test his FL through what he called his "synthetic" deductions. Consequently, Weber's next task was "to synthetically deduce [from the fundamental law] a system of consequences", in order to test them against the known laws of electrodynamics. He diligently deduced[35] the law of electrodynamic action of a closed circuit on a current element; of a magnet on a current element (Gauss's law); the law of Volta-induction of a closed current on an element of moving conductor; the magneto induction of a magnet on an element of moving conductor; the induction on a stationary conductor by the approach or separation of a constant current element; the induction of one conductor due to the variation of intensity in a neighbouring one; and, finally, a general law of Volta-induction.

In the years 1848-55 Weber extended his FL to other types of measurements. One of his main tasks was the measurement of resistance in absolute units. In a work[36] published in 1851, he showed that when absolute magnetic units for current intensity (Weber: "the absolute unit of current in the magnetic system ") and absolute units for electromotive force are determined, it is possible to define absolute units of resistance as well.

Weber's program of precise definitions and measurement was taken up again in 1852, when he studied[37] how his absolute systems of units could be extended to comprise the resistance of conductors through Ohm's law. In his determination of units for potential, he used the electromagnetic induction law and the determination of the unit of magnetic flux variation, a procedure which echoed Gauss's work on absolute magnetic units.[38]

Weber stressed that only in a system of absolute units could the multiplicative proportionality constants of Ohm's law and of Coulomb's law for the electrostatic force be pure numbers (i.e., dimensionless) and have unit value. He wanted to prove that in his

FL, another constant appeared which did not have unit value. He soon discovered that in order to have a quantitative base for his definition of a complete system of electric quantities, a numerical knowledge of this constant was indispensable.

Since the law was for him a fundamental one, a fundamental constant somehow related to the ratio m above should have fundamental significance, i.e., it should be a fundamental constant independent of the choice of units and systems.

Weber's research program in the following ten years was mainly aimed at determining and measuring this constant.

In his 1846 paper[39], he had reached an initial conclusion about the existence of this constant. Introducing an as yet unknown velocity v, and writing:

$$\frac{1}{2} v^2 = \frac{1}{2}(\frac{dr}{dt})^2 - r \frac{d^2r}{dt^2} ,$$

he found that his FL assumed the form:

$$ee'(1 - v^2 \frac{a^2}{16}) .$$

The factor reducing the Coulombian force was, therefore, proportional to the square of this velocity v. He called v "the reduced relative velocity" ("reducierte relative Geschwindigkeit").

In 1848, he distinguished[40] the (units-dependent) number a (in his expression (1) of the FL) from the reduced relative velocity, which he now indicated as V, by remarking that, with a suitable choice of unit of time and space, such that $t^* = 4\,t$, and $R = m\,r$, the FL of force between two masses of either signs (formula (2)) could be rewritten:[41]

$$EE'/R^2 \, [1 - (dR/dt^*)^2 + 2R \, d2R/dt^{*2}] \, .$$

When the two masses are in the same point and the relative velocity is constant, $R = 0$, $dR/dt^* = V$, and so:

$$EE'/R^2 \, (1 - V^2) \, .$$

The reduced relative velocity V was thus defined as the constant whose squared value represented the factor for reducing the Coulomb force, when the two masses have speed V. This constant accordingly had fundamental significance. Finally, in an extended series of experiments on the determination of the absolute unit of resistance published in 1852,[42] he argued that, posing: $c^2 = 4/m^2$, his FL (formula (2) above) could be written in a neater form:

$$\frac{ee'}{r^2}\{ 1 - (1/c^2) \, (\tfrac{dr}{dt})^2 - (1/c^2) \, 2r \, \tfrac{d^2r}{dt^2} \} \, .$$

If dr/dt is constant:

$$\frac{ee'}{r^2}\{ 1 - (1/c^2) \, (\tfrac{dr}{dt})^2 \} \, .$$

Thereafter, c is defined: "[c is] that constant value assumed by the relative velocity dr/dt when two electric masses do not exert forces on each other."[43]

This definition of c determined solely by the condition that the force disappears makes it fully independent, by definition, from any conventional ratio of units.

In Section IV of his 1852 paper, Weber underlined the connection between the derivation of his FL and the foundation of complete systems of absolute units for electric and magnetic quantities. He emphasized the point that in line of principles the determination of a complete system of units for electrodynamics "without any relation with the magnetic measurements" was thus achieved.[44] However, he added: " without knowing the velocity c, the measured intensity of current, of electromotive force and of resistance cannot be related to known magnetic measure".[45]

Clearly, in the absence of a known value for c, all the effects generated by the motions of a known (in electrostatic measure) electric charge could not be predicted through his FL:

The velocity c....is up to now not yet determined and this is the reason why the above measures are not utilizable in the practical applications of Electrodynamics, because, without a knowledge of the velocity c, reduction to the known measures of mechanics of the measured *current intensity, electromotive force and resistance cannot be completed.*[46] [Italics in the original].

The proof that c was proportional to a ratio of units was given many times and with different approaches in Weber's papers. Let us summarize in a table these proportionality relations as they are reported in Weber and Kohlrausch's 1856 paper[47] (I use ® to denote ratios of different units):

Ratio of units for current	$c = g$ ®
®1 = magnetic/mechanical	$c = 2 \sqrt{2}$ ®1
®2 = electrodyn/mechanical	$c = 4$ ®2
®3 = electrolytic/mechanical	$c = 3 \sqrt{2}\ 106$ ®3

(Table A)

This table clearly show that Weber's c and ⓡ are measures of numerically different quantities related by Weber's FL. As proved by Weber's theory above, these are also conceptually different quantities, because c is the measure of a constant velocity having a precisely definite physical meaning, while ⓡ is a variable ratio of units.

The constant c could be determined by a measurement of the dynamic effect of a moving charged body. It seems that Rudolf Kohlrausch, Weber's collaborator in the measurement of c initially supported an experiment of this type.[48] However, Weber and Kohlrausch thought that, even if large charges were moved, the velocity necessary to obtain measurable effects would have been very large[49]. Therefore, they adopted a more practical method.

This method consisted[50] in the measurement (through the tangent galvanometer used ballistically) of the magnetic force produced by the discharge of a known electric charge. The instantaneous effects on the moving magnetic needle of this instrument (i.e., the needle's initial angular velocity) depended in this case on the speed of the charge's transfer. The duration of the needle's deflection varied reciprocally with this speed and so the time-integrated effects measured by the complete deflection would, therefore, be independent of the discharge speed. This method was practical because there were instruments which could be used to measure time-integrated effects including Weber's electrodynamometer and tangent galvanometer.

The instrumental operations for measuring c demanded particular shrewdness because many of the quantities involved needed a precision never attained before and operations with auxiliary instruments were also necessary. A short description[51] of the main features of the experiment follows.

A capacitor (Leyden jar) is charged, and the charge Q on one of its two plates is measured in electrostatic units.[52] The same capacitor is then discharged through a coil, and the time-integrated current which flows through the coil is measured through the deflection of the tangent galvanometer.[53] In order to measure in magnetic units Q, the charge of the jar, the same deflection of the tangent galvanometer was produced through a constant unit current (in em units) flowing for a short time τ. (I use throughout the letter Q in place of Weber's E, for avoiding confusion between Weber's notation and the E used by Maxwell, as reported in the following chapter). Due to Weber's conception of current, $Q/2$ electrostatic units flowed in the positive direction in each wire. Hence $Q/2\tau$ measured the ratio between magnetic and mechanical units of charge (or equivalently of current).

Weber and Kohlrausch found that the measured value was: $Q/2\tau = 155370 \times 10^6$, i. e., 155370×10^6 mm/sec, corresponding to the quantity of electricity of one sign (in es units) carried by a unit (in em units) current flowing for one second in a wire.

Clearly, $Q/2\tau = \circledR 1$ is the units ratio in table A above, and from the relation, in the same table, $c = 2\sqrt{2}\circledR 1$, Weber and Kohlrausch computed the value of c:

$c = 439450 \times 10^6$ units of length per second (Weber's unit of length = 1 mm).

Weber remarked that in his FL, c^2 represented the ratio between the electrostatic (charge-charge force) and the electrodynamic (current-current force) force in mechanical units. Therefore, his theory: "explains why the electrodynamic interaction in electrical units... always appears to be infinitesimally small in comparison with the electrostatic interaction ee'/r^2 ; so that, in most cases, the former remains insignificant only if, as in a galvanic current, the electrostatic force is completely eliminated, on account of the neutralization of positive and negative masses".[54]

The following often quoted 1856 remark is of interest in connection with Weber's views on a possible identification between his c and the velocity of light:

> In all the laws in which the constant c occurred, it appeared in the denominator of the ratio between it and the velocity with which the bodies move with respect to one another. It is then of practical interest that all effective velocities which we know of, even those of celestial bodies, can be considered as vanishingly small in relation to c. *The only velocity we know of that approaches c, namely that of the propagation of light, is not a velocity with which bodies in effect move relative to one another* (Italics are mine).[55]

Notice that one of the difficulties for the identification presented by Weber, i.e., the different physical nature of the two velocities, was precisely the one which Maxwell tried to overcome in his exploitation of Weber's results for his electromagnetic theory of light.

In 1857 Gustav Robert Kirchhoff found[56] that his theory of propagation of currents in wires led to the result that waves of current existed and propagated with the velocity $c/\sqrt{2}$, closer to the speed of light. Later on in 1864, Weber proved[57] that $c/\sqrt{2}$ was the velocity of current waves but only in limiting conditions such as those occurring in long and thin conductors of negligible resistance. He concluded[58] that, concerning this velocity, "values very close to such a limit are very rare".

Weber's and Kirchhoff's context of ideas, embedded as it was in a particle-based approach to electrodynamics permitted a mere numerical quasi-equality between c and the light velocity to have little significance. Consider in this regard the role that Weber and Kirchhoff attributed to c in their proposal of a generalized metrological program, based on two fundamental constants.

In their 1856 paper,[59] they proposed that both Newton's gravitational law and the fundamental law of electric action be assumed as fundamental laws of nature (*Grundgesetze der Natur*). This allowed the elimination of mass among the fundamental quantities, because:

> ...all other measurements can be simply derived from the two fundamental measures of space and time, .. the measure of mass can be reduced to the latter though the gravitational fundamental law.

A further step in the reduction of the fundamental units could be made by taking c as a unit:

> One can assume as a unit of time the time spent by two electric masses for approaching or departing of one unit of space, when, according to the same law, they move at such a speed as to exert no reciprocal action.[60]

In conclusion, let us summarize the main conceptual features of Weber's theory presented in the foregoing pages:

a) Through his measurement of c, a constant velocity in his fundamental low of force, Weber solved the long known problem of the ratio between the force exerted by a given quantity of frictional electricity standing in a condenser and the force it exerted when flowing in a wire, a problem that intrigued Faraday in his researches.

b) In Weber's theory, the characteristic velocity c and ®, the ratio of units, represented numerically different measures of conceptually different quantities. In fact, c is the constant velocity of motion of two electric 'masses' when they exert no mutual actions while ® is a variable ratio of units.

c) Weber's c derived its meaning from Weber's fundamental law and from Weber's convective conception of current, a conceptions that was a necessary assumption for the demonstration of Weber's law.

d) Weber's fundamental law represented a generalization of Ampère's and Faraday's laws. These laws are then valid independently of the validity of Weber's law.

e) Weber and Kohlraush measured ⓡ1 through instrumental operations based on Ampère's law; therefore (by d)), ⓡ1 was measured independently [61] of Weber's fundamental law.

1.5. Comments

No wonder that, because of his making the above committing statements, Ampère's derivation of the elementary law met all sorts of criticism, and, at the same time, many expressions of praise and admiration.[62]

Among Ampère's critics the first, and one of the most outright, was the same Wilhelm Weber that extended his research. In 1845 he wrote that Ampère did not correctly deduce his law from experiments, because he overlooked all problems concerning the sensitivity of his apparatus and, consequently, the precision of his measurements.[63] More specifically, Weber's criticism concerned the fact that in Ampère's "zero-point measurements" a determination of the error is indispensable. Since Ampère obtained the connections between the fixed and the movable portions of circuits through pots and channels filled with mercury, friction and other unwanted interactions must have interfered with the establishment of the positions of equilibrium.

At the beginning of this century, another subtle critic of Ampère's improper conclusions of an inductive inference from factual observations was Pierre Duhem.[64] Duhem objected that "les faits d'expérience, pris dans leur brutalité native ne sauraient servir au raisonnement mathematique ; pour alimenter ce raisonnement, ils doivent être transformés et mis sous forme symbôlique".

According to Duhem, Ampère achieved this symbolic transformation when he implicitly formulated the following hypotheses:[65]

E. Forces exerted by closed circuits are the superposition of forces between all the possible pairs of elements in both circuits.

F. Forces are central and they obey the principle of action and reaction.

Taking our stand on this formidable problem, no less than that of the factual foundation of an empirical science, clearly those statements that Weber and Duhem considered hypotheses, were for Ampère mechanical principles, constitutive foundational axioms; on these foundations he constructed his theory. The notion that forces were central was for him an axiom constitutive of the force concept:

Je n'ai fait aucune recherche sur la cause même qu'on peut assigner à ces forces, bien convaincu que toute recherche de ce genre doit être précédée de la connaissance purement expérimentale des lois, et de la détermination, uniquement déduite de ces lois, de la valeur des forces *élémentaires dont la direction est nécessairement celle de la droite menée par les points matériels entre les quels elles s'exercent* [my italics]. [66]

Thus, the axiom of central forces was for him a necessary premise for the lawfulness of any balance-type experiment. Ampère's inductive inference receives its support from this axiom: so far as Newton's law *s'accorde avec les faits*, Ampère's law is also to be considered *uniquement dèduit de l'expérience*.

For some aspects, Weber's *Electrodynamische Massbestimmungen*, the work examined above, being a long and detailed study on the theory of interactions between moving charges, represented a completion of Ampère's ideas.

Weber had in common with Ampère the conception that electrodynamics was a more basic expression of electrical phenomena than electromagnetism. He pursued a step further the Amperian mechanical scheme of electrodynamics by assuming that the Amperian forces between elements of current were the manifestation of deeper forces between particles of electricity acting at a distance. He believed that the law of these forces, his fundamental law, could be inferred, not deduced, from the experimental laws of Coulomb and Ampère.

Among others, Weber's emphasising Ampère's view that galvanism was more fundamental than magnetism allowed a definition of electrodynamic and electromagnetic units of current intensity independently of Gauss's method of current-magnet interaction, thus providing a simpler basis for the absolute measure of electric current in terms of fundamental mechanical units. In his determination in the *Treatise* of the electromagnetic units of current intensity, Maxwell followed Weber's approach.

Historically, Gauss's and Weber's introduction of absolute systems of units permitted physical laws to be expressed in richer forms of mathematical equations than the consideration of proportionality relations and purely local numbers made possible, the form in which physics laws were symbolized in the writings of Coulomb, Fresnel and others physicists of the beginning of the nineteenth century.

I have shown above the interconnection between theory, measures and instrumental operations which resulted in this new form of representing the laws of physics. Quantitative prediction was one of the most important issues involved.[67]

As shown in the foregoing discussion of his discovery of c, Weber profited from the use in his equations of symbols representing numbers, without otherwise explicitly introducing a theory of dimensions (as Maxwell did). Since numbers have null dimensions, Maxwell's innovation of dimensional quantities can be historically considered as a generalization of Weber's initial position.

The discovery of c, fundamentally important for a FL of electrodynamics, was also interpreted by Weber as having a metrological role in new formulations of the mechanical units of space and time and of mechanics itself. Weber interpreted the above discovery as an indication of a new connection between space and time units, which allow the fundamental mechanical quantities and units to be reduced from three to two .

CHAPTER 2

A SURVEY OF THEORIES OF UNITS AND DIMENSIONS IN NINETEENTH-CENTURY PHYSICS

Metrological theories and theories of dimensions are tightly connected with the nineteenth century history of electrodynamics. Their significance is better understood when these theories are situated in an historical perspective.

When Gauss and Weber decided to use physical laws tto define a system of units, rather than transcribing these laws in the form of simple proportionality-relations among quantities measured in arbitrary units, physical laws could be conceived and written as analytical equations whose symbols represented rational numbers. A great innovation was thus achieved in the physical sciences.

Gauss and Weber did not deal explicitly with dimensions. The explicit introduction of dimensional arguments into his *Theorie Analytique* was the task of Joseph Fourier in 1822.[1] Invariance of the equations under change of units is Fourier's main concern in introducing dimensions:[2]

Il faut maintenant remarquer que chaque grandeur indéterminée ou constante à une dimension qui lui est propre et que les termes d'une même équation ne pourraient pas être comparés, s'ils n'avaient point le meme exposant de dimension. Nous avons introduit cette considération dans la Théorie de la chaleur pour rendre nos définitions plus fixes et servir à vérifier le calcul; elle dérive des notions primordiales sur les quantités: c'est pour cette raison que, dans la Géometrie et dans la Mécanique, elle équivaut aux lemmes fondamentaux que les Grecs nous ont laissés sans démonstration...

Dans la théorie analytique de la chaleur, toute équation (E) exprime une relation nécessaire entre des grandeurs subsistantes x, t, v, c, h, k. Cette relation ne dépend point du choix de l'unité de longuer, qui de sa nature est contingent; c'est-à-dire que, si l'on prenait une unité différent pour measurer les dimensions linéaires, l'équation (E) serait encore la même.

In his *Théorie Analytique,* only dimensions of length, time and temperature are involved in the equations. Maxwell considered Fourier as his source in his 1863 *Report* and in his *Treatise,*[3] where he introduced the modern notation for dimensions, using capital letters in square brackets. In this work, letters in the equations symbolised quantities (i.e., a number times a unit), not simple numbers as in Weber.[4]

William Thomson is in good company with Maxwell in identifying dimensions of a quantity in a given system with the quantity itself (This I call Maxwell's transitional principle). In Thomson's hyperbolic mood:

I suppose almost everyone present would think it simple idiocy if I went to say that the weight of that piece of chalk is the fourth power of seven or eight yards for hour; yet it would be perfectly good sense.

This type of identification is what Maxwell proposed in his 1863 *Report* and in his *Treatise.* Some Maxwellians (e.g., J. E. H. Gordon, Everett, etc.) followed Maxwell's views. William Kingdom Clifford, on the other hand, took a more articulated position in 1878, underlying the conventional character of the new symbolisation of dimensions and justifying its convenience in the calculation of the change of units; however, opposing Thomson, he warned[5] that this convenience could be a cause for "nonsense" if the meaning of dimensions was unduly extended:

$[V] = [L]/[T]$...Here the word per has been replaced by the sign for divided by.

Now it is nonsense to say that a unit of velocity is a unit of length divided by a unit of time in the ordinary sense of the words. But we find it convenient to give a new meaning to the words "divided by" and to the symbols which shortly expresses them... this convenience is made manifest when we have to change from one unit to another....

In 1982 Rudolf Clausius presented[6] two lines of criticism to Maxwell's dimensional theories. The first line concerned a particular remark on the attribution of dimensions to the product [pC] in Maxwell's *Treatise* (Part IV, Chapter X). This attribution is considered by Clausius to be a flaw in Maxwell's derivation, which produced extended consequences in the attribution of wrong dimensions to the magnetic quantities.[7] He proposed a different line of approach in order to obtain the derivation of dimensional relations. The results are equal to M.'s but for the dimension of the quantity of magnetic pole, in the electrostatic system. Clausius went on to examine the problem of Weber's factor and he explicitly criticised Maxwell's statement that units of the same quantity belonging to two different systems[8] can be related through a dimensional ratio. He considered this statement mathematically incorrect and he proposed a different method which allows the relation to exist only between units of different quantities in the same system. The conversion factor between units is, in Clausius, a pure number.

Max Planck's admission of a plurality of systems was born from his conviction that the idea of dimension applies only relatively to a given system and that it is "nonsense" to search for the " real " dimension of a quantity. Consequently, he deprived Maxwell's two absolute systems from any privilege and considered them as part of a set of other possible systems. This conclusion wiped out a possible objection to the plurality of dimensions of a quantity but, at the same time, deprived Maxwell's two-systems theory of its label of being "the only scientific" theory, and dismissed Maxwell's argument in favour of Weber's equality:[9]

The fact that when a definitive physical quantity is measured in two different systems of units it has not only different numerical values, but also different dimensions, has often been interpreted as an inconsistency that demands explanation, and has given rise to the question of the real dimensions of a physical quantity. After the above discussion it is clear that the question has no more sense then inquiring into the real name of an object.

Giovanni Giorgi (1871-1950) was convinced in 1901 that Maxwell's electromagnetic absolute units were practivally useless because of their size, and that their shortcomings were not reparable by appropriately redefining the fundamental units, but required the introduction of a fourth unit. He urged the abandonment of the three-unit system, which had been inspired, according to him, by Gauss's and Weber's mechanistic approach to electrodynamics. Giorgi proposed the well-known four-unit system,[10] using the Ohm as the fourth unit for electric resistance. What he objected[11] to in Maxwell's theory was that, in Coulomb's laws for electric and magnetic charges, the constants could not bereduced to one, because they were not numerical coefficients but "physical constants", an index of " the capacity of space of being charged with energy" (in our modern view: the three-unit system is overdetermined).

In the 1933, the American Committee of Physicists and Electricians proposed that the Special Committee of the Electrotechnical Commission adopt Giorgi's four-unit system, and the Commission in October 1933 approved the proposal. In 1935 Arnold Sommerfeld decided[12] in favour of the four-unit system:

The orthodox number three, which is at the base of the so-called absolute system of measurement, could be considered as mandatory as long as one hoped to derive electricity from mechanics. This time is now over. One exerts violence against electromagnetic quantities if one forces them, in the Procuste's bed of the three units; it can be shown that they are at ease in the four-unit system.

Sommerfeld harshly criticised the three-unit system in his celebrated *Electrodynamics*,[13] where he warned against what he considered the ruinous pedagogical effects of Maxwell's two absolute systems:

We have frightened generations of students with these two sets of values for charges and field strength.

He also quoted among "so many other clarifications in the question of units", an irreverent (to Maxwell's approach) example by J. Wallott, in which a fundamental velocity is made to appear in acoustics through an appropriate selection of constants and units.

CHAPTER 3

A HISTORICAL ROLE FOR DIMENSIONAL ANALYSIS IN MAXWELL'S ELECTROMAGNETIC THEORY OF LIGHT

3. 1. Maxwell Transforms Electrodynamics into Electromagnetism

Maxwell's contributions to physics have been extensively scrutinised by historians in the last decade, but certain aspects of his work, however, are still partially unexplored. Many of Maxwell's historians have perhaps favoured those parts of Maxwell's work which are, more or less, related to our modern theory. The consideration of some outmoded or controversial parts of his theories, such as the ones dealt with in this paper, will contribute, I hope, to a better understanding of the historical situation of Maxwell's electromagnetism.

An example of the controversial aspects in Maxwell's theories is, in my opinion, Maxwell's electromagnetic theory of light and, in particular, the metrological and dimensional theories which represent a large part of its supporting evidence. These theories areused by Maxwell in order to prove the equality between Weber's conversion factor and the velocity of light, one of the main pieces of evidence in favor of his optical theory.

It is known that Maxwell brought into his electromagnetic theory of light concepts and data which were obtained in the 1850's by Weber. Maxwell's conclusion was that Weber's conversion factor between electrical units, which in Weber's theory corresponded to the relative velocity of motion of electric particles, represented for him the velocity of electromagnetic waves and of light in an electric ether. This conclusion was a momentous one because it transformed a relative velocity of motion into an absolute velocity of waves.

The conceptual difficulties presented by this transformation are documented in Maxwell's papers and specifically in the many routes he followed in the course of his life in order to reach a satisfactory demonstration of it.

As is known, Maxwell used arguments of various kinds to prove the electromagnetic theory of light, of which the most important one is the demonstration that electromagnetic waves propagate in the ethereal medium with the velocity of light. This demonstration is presented in different forms along with the development of Maxwell's electromagnetic theory, since the first paper he devoted to this problem in 1862, until his 1873 masterwork: *A Treatise on Electricity and Magnetism*.

In all of these works Maxwell is faced with the problem of checking a theoretical value of the velocity of the electromagnetic waves, calculated by his electromagnetic field theory, against the velocity of light, known to him because of theoptical measurements made by Fizeau (1849) and Foucault (1862), and also because of the measured aberration of light. An independent measurement of the velocity of the electromagnetic waves (heretofore: EMW) was, in Maxwell's time, beyond the reach of experimental technique. Heinrich Hertz, as is known, experimented and measured the velocity in question eight years after Maxwell's death. Maxwell however succeeded just the same in measuring indirectly the EMW velocity and found that it was approximately equal to the velocity of light.

In this paper I am primarily concerned with Maxwell's indirect procedure of using Weber's factor to measure the EMW velocity and with the difficulties he faced in reaching a satisfactory demonstration of this procedure. These difficulties are mainly due to the fact that the existence of Weber's factor, i.e., of a ratio between values obtained in measuring the same charge in electrostatic and in electromagnetic units, was a direct consequence of those parts of Weber's theory, i.e., the fundamental action-at-a-distance law and the convective conception of currents, which Maxwell did not accept. Due to this, Maxwell introduced into his theory Weber's factor as a consequence of dimensional equations.

In his theory,[1] Weber defined various systems of units for measuring the electric and magnetic quantities, which he called "absolute systems", and introduced the concept of a characteristic velocity of motion for electric particles. In 1856 he measured it as a "conversion factor" between different units of the same quantity, thanks to the aid of his colleague, the experimentalist Rudolf Kohlrausch. Together they found[2] this factor (henceforth named Weber's factor) to be approximately numerically equal to the velocity of light, but the two German physicists did not attach any significance to this result as a possible hint about an electric theory of light.

Maxwell developed various theories in order to solve the problem of proving that Weber's factor was in fact the velocity of EMW and of light. For the understanding of what follows, it may be helpful to realise that the various solutions given by Maxwell to the problem above can be considered as consisting of two distinct parts which I label as part A and part B.

Part A connects the velocity of the electromagnetic waves to the ethereal constant k and µ, the " dielectric capacity" and the " magnetic permeability" of the electromagnetic ether. For this part Maxwell's own theory sufficed.

As regards part B, it connects the same ethereal constants to

Weber's factor and the velocity of light. Part B presented more diffculties in as much as Weber's theory was an action-at-a-distance theory, and the ethereal constants were foreign to it.

A correlation between parts A and B was required to prove that the velocity of the electromagnetic waves was equal to Weber's ratio and the velocity of light. This correlation constituted in fact the above mentioned indirect measurement of EMW velocity.

The development of the complex set of conceptions and theories, which largely deal with part A, have been adequately analyzed in recent years by historians [3]; however, part B has until now escaped their attention, although I think that this part also offers interesting clues.[4] In fact, in this part Maxwell is confronted with the problem of transforming Weber's factor fromone concerned with a particle's velocity, as it appeared in Weber, to one decribing a wave's velocity. The ethereal constants played the role of midwives in the transformation.

Maxwell's difficulties with this problem, and his struggle to find arguments in support of the above correlation, are significantly documented in his various attempts to connect Weber's factor to the constants of the ether, a necessary prerequisite for concluding that the same factor represented the velocity of electromagnetic waves in ether.

3.2. In order to Adopt Weber's Result, Maxwell Resorts to the Elastic Theories of Optics

To my knowledge, Maxwell's first quotation of Weber's theory is in a letter to William Thomson dated May 15, 1855;[5] following Thomson's suggestion, Maxwell read Weber's *Elektrodynamische Maas-*

bestimmungen and his comment was:

I have been examining his (i.e.Weber's) mode of connecting electrostatics with electrodynamics, induction etc., I confess I like it not at first... but I suppose the rest of his views are founded on experiments which are trustworthy as well as elaborate .

Six years after Maxwell's first approach to Weber's work one can read a rather extended comment on his involvement with a new theory of light: in a letter [6] mailed from London, on December 10, 1861, he described to Thomson his particular ether's model of particles and cells, adding that he "calculated the relation between the forces and the displacement on the supposition that the celles are spherical and that their cubic and linear elasticities are connected as in a 'perfect' solid". In the same letter, he hinted at Weber's value of the statical measure of a unit of electric current, from which he deduced the relation between the elasticity and density of the cells, and the velocity of transverse ondulations. He then advanced the bold hypothesis :

the magnetic and luminiferous media are identical and... Weber's number is really, as it appears to be, one half the velocity of light in millimeters per second.[7]

Maxwell's initial impact on Weber's metrology and the velocity of light were destined to a great "seguito".

The story continues in Maxwell's paper, "On Physical Lines of Force", published in the *Philosophical Magazine*.[8] The first two parts, dealing mainly with a field theory of electrostatics, Faraday's induction and Ampere's forces, were published around March 1961. In Part 3, " The Theory of Molecular Vortices Applied to Statical Electricity", published in the January and February 1862 issue of the same magazine, Maxwell presented[9] his first version of the electromagnetic theory of light, and for the first time, in a published paper, he referred to the experimental work of Weber and R. Kohlrausch.[10]

In order to fully grasp Maxwell's conception at this stage, one should consider that his work on the theory of light had been preceded by a half century of research and publications on the elastic theories of an optical ether. Augustin Fresnel and Augustin Luois Chauchy, among others, succeeded in showing how some complicated effects of christalline optics could be explained by apt elastic hypothesis and mathematical analysis. George Green and Gabriel Stockes in England had pursued the methods of mathematical-physics to work out elastic theories of an optical ether. Maxwell himself contributed[11] to elasticity in one of his first scientific papers.

This background is relevant for an assessment of Maxwell's initial approach to Weber's factor, explaining why he accepted that, in his first identification of this factor with the velocity of light, the elastic theory could still have a role. In fact, in part 3 of his paper he presented[12] a detailed hypothesis on the elastic properties of an "electric ether", through which the relation between the "electric displacement" and the "electric force" are deduced.

Analogy plays an important role in the deduction: each quantity is endowed with a double referent, one referring to the elastic and the other to the electric theory. For example: the quantity E, "a coefficient dependent on the nature of the dielectric", connects the "electromotive force" R to the "electric displacement" h, in the equation:

$$R = -4E^2 h.$$

The same equation possesses, however, a counterpart in elasticity, representing Hook's law of force, a special case of a strain-stress relation.

Similarly, the magnetic quantity μ, the "coefficient of magnetic induction", also possesses a counterpart: the "density of the matter of the vortices". Once the elastic roles of E^2 and μ are assigned, the elastic law $V=\sqrt{E^2/\mu}$ correctly represents the velocity of propagation of a wave in an elastic medium and, assuming that the "density of the matter of the vortices" μ is unitary, one has:

$$V = E.$$

In his working out of part A, sect. 3.1 above, Maxwell is thus helped by special hypotheses on the elasticity of the ether.

He has now to show that this velocity equalled Weber's factor, i.e., part B. From the equation $R = -4E^2h$ and a theorem on the strain and energy in an elastic solid, he deduced [13] Coulomb's law, in the form: $F = -E^2 e_1 e_2/r^2$ (second form). Because the units are, according to Maxwell, electromagnetic units, a comparison with the same law in the usual form:

$$\frac{n_1 n_2}{r^2}$$

— i.e. "measured statically"—allows one to attribute to E the meaning of a conversion factor between "dynamic and electrostatic units". Provided dynamic units are identified with Weber's electromagnetic units, the above factor is Weber's factor. In short, magnetic disturbances, predicted in theory, propagate with a velocity which is equal to Weber's factor and the velocity of light.

After that, Maxwell [14] announced his first prudent statement on his new theory of light:

I have deduced from this result the relation between the statical and the dynamical measures of electricity, and have shown, by a comparison of the electrodynamic experiments of M.M. Kohlrausch and Weber with the velocity of light as found by M. Fizeau, that the elasticity of the magnetic medium in air is the same of that of the luminiferous medium, if the two coextensive, and equally elastic media are not rather one medium.

The strength of Maxwell's argument is the partial analogy between electromagnetic phenomena, on the one hand, and elastic behavior of solids on the other, a rather well known and accepted analogy for the British electric scientists of the middle of the century, ranging from Faraday's analogies on the behaviour of currents in cables to the mathematical elaboration of elastic analogies by William Thomson in the fifties.

Its weakness results from its attribution of specific elastic properties to ether. Besides, its failure to specify a definite system, the attribution of units is ambiguous, because there is no precise reasons why dynamic units are to be identified with Weber's electrodynamic units.

The above usage of units should have appeared unsatisfactory to Maxwell since his next step was the development of a metrological theory of two clearly defined absolute systems of measurement.

3. 3. Metrology makes its Entry on Maxwell's Optical Stage

In 1862, The British Association for the Advancement of Science, founded in 1831, appointed a Committee on Electrical Standards, to decide on the best system of electrical units. In the same year, Maxwell joined the committee which comprised among its members Charles Wheatstone, William Thomson, Franklin. Jenkin, C. W.

Siemens, Belfour Stewart and James P. Joule.[15]

At that time the first two Parts of his " On Physical Lines of Force" had been recently published. In the January and February 1862 issues of *Philosophical Magazine*, Maxwell presented his first version of an electromagnetic theory of light.

We have a clear indication of Maxwell's engagement in 1862-63 in the activities of the Committee on Electrical Standards in an account given by one of Maxwell's students, W. D. Niven:

In 1862-63 [Maxwell] took a prominent part in the experiments organized by a Committee of the British Association for the determination of electrical resistance in absolute measure and for placing electrical measurements on a satisfactory basis. In the experiments which were conducted at King's College upon a plan due to Sir W. Thomson, two long series of measurements were taken in successive years.[16]

Maxwell and Thomson's activity had a determinant role in orienting the committee's work towards a metrological program of a high scientific level, inspired by Weber's absolute systems of electric and magnetic units. A perusal into the annual Reports of the British Association for the same period shows[17] Maxwell and Thomson's success in shifting the committee's program from the definition and construction of "etalons" (an engineering response to the need for an interpersonal measurement of electric and magnetic units) to the more difficult but highly significant determination, by laborious laboratory measurements, of units in each of the two absolute systems.[18] In Maxwell's (and Fleming Jenkin's) Appendix C to the *Report* entitled "On the Elementary Relations between Electric Measurements" (1863), the conception of two complete absolute systems of electrostatic and electromagnetic units is presented [19] as the only one consistent with the present knowledge of electromagnetic phenomena and of their connection with the mechanical measurement of space, time and mass. The two authors remark that the discovery in 1841 of Joule's law on the thermal effect of an electric current made possible the completion of the electrostatic system.

In Appendix C to the same 1863 *Report*,[20] Maxwell dealt explicitly with dimensions,[21] mentioning Fourier[22] as his source :

> Dimensions of Derived Units. - Every measurement of which we have to speak involves as factors measurement of space, time and mass only; but these measurements enter sometime at one power, and sometimes at another. In passing from one set of foundamental units to another, and for other purposes, it is useful to know at what power each of these fundamental measurements enters into the derived measures. Thus the value of a force is directly proportional to a length and mass, but inversely proportional to the square of a time. This is expressed by saying that the dimensions of a force are LM/T.

Maxwell refers also to the French scientist in his *Treatise*.

3.4. The Transitional Principle

What deserves our attention is the link Maxwell establishes between the dimension of a quantity and its physical attributes. He assumes that, if a quantity X has the dimension of a velocity, this quantity *is* a velocity. This assumption represents a transition from dimensional to physical proprieties, a transition now denied in modern physics. Let us name it the *transitional principle*.

One example of its application concerns[23] the resistance R of a conducting wire. By Faraday's induction and Ohm's law, this resistance is expressed as:

$$R = \sqrt{\frac{VSL}{C}} \; ; \qquad (8)$$

V indicates the velocity of motion of the conductor of length L, traversed by a current C in a magnetic field of intensity S.

Let us follow Maxwell's argument in support of the above

principle:

One curious consequence of these considerations is, that the resistance of a conductor in absolute measure is really expressed by a velocity; for, by equation (8)[24], when SL = C we have R = V, that is to say, the resistance of a conductor may be expressed or defined as equal to the velocity with which it must move, if placed in the conditions described, in order to generate a current equal to the product of the length of the conductor into the intensity of the magnetic field; *or more simply, the resistance of a circuit is the velocity with which the conductor of unit length must move across a magnetic field of unit intensity in order to generate a unit current in the circuit* (italics are mine).

In short, if electric resistance *is dimensionally* a velocity, it *is physically* a velocity. The *transitional principle*, once established, will legitimately allow the transition from any dimensional to any physical property. Maxwell introduced this principle in the above-cited passage, and will emphasise and expand on it in *A Treatise in Electricity and Magnetism*.

The *Report* for 1863 ends with the authors' announcement that they will include a measurement of Weber's factor in the plans for the Committee is future works.

In a letter[25] from Maxwell to Gabriel Stokes dated 15 October 1864, we have a first-hand indication that Maxwell thought he had found a way of by-passing his approach to the problem adopted in *"On Pysical Lines"*:

I have now got materials for calculating the velocity of transmission of a magnetic disturbance through air founded on experimental evidence, without any hypothesis about the structure of the medium or any mechanical explanation of electricity and magnetism.

In his letter Maxwell mentioned the problem of the velocity of " slow " and "rapid" disturbances and so concluded the passage: "We are devising methods to determine this velocity = electromagnetic:electrostatic units of electricity..."

In the same letter he referred both to his and Jenkin's experiment on the ways of measuring the "capacity of a conductor both ways" and to "a plan of a direct equilibrium between an electromagnetic repulsion and electrostatic attraction"; this is a clear hint that the planned measurement of Weber's factor is the one described in his "Note"(1868).

It is then reasonable[26] to argue that in the passage above Maxwell's "experimental evidence" referred to the more direct link between the measurement of Weber's factor and the velocity of the waves. He had established this connection through the metrology of the two systems, one that he will exploit in his essay "A Dynamical Theory of the Electromagnetic Field".

3.5. Elimination of Hypotheses and Experimental Evidence in a "Dynamical Theory"

In his 1864 memoir "A Dynamical Theory of the Electromagnetic Field", [27] Maxwell derived the second form of Coulomb's law from a field energy expression (deduced from Hooke's law) and from Gauss's law:[28]

$$\{Energy\} = \int Pdf = 1/2 \; Pf \; ;$$

$$div \, f = -e \, , \, P = kf$$

P is the x component of the Electromotive force, and f the x component of the electric displacement; e is the density of charge.

From the equation above, Coulomb's law is deduced in the form:

$$\frac{K \, e_1 e_2}{4\pi \, r^2} = \frac{v^2 \, e_1 e_2}{r^2} \; ,$$

e_1, e_2 are quantities of electric charges in electromagnetic units;
k is the "dielect capacity of ether": $k = 4\pi v^2$.

Coulomb's law in the second form is thus " expressed in terms of the Electromagnetic System of measurement which is founded on the mechanical action between currents". By comparison[29] with the same law in electrostatic units:

$$\frac{n_1 n_2}{r^2},$$

V in the first equation above, is Weber's factor.
All that accounts for part B.

Regarding part A, the other important achievement is the derivation[30] on purely electromagnetic grounds of a D'Alembert-type equation for the propagation of magnetic field (k enters as usual as a factor in the displacement current term). This allows us to deduce V the velocity of magnetic waves from the equation, without any recourse to a spurious elastic ingredient.

The propagation velocity is:

$$V = +\sqrt{\frac{k}{4\pi\mu}}.$$

Since in air, $\mu = 1$ electromagnetic units, the propagation velocity of the waves equals Weber's factor:

$$V = v.$$

In the passage above, Coulomb's law is derived dynamically, by representing electric forces as stresses in an elastic medium by Hooke's law. The burden of the proof rests on Gauss's law and on the expression for elastic energy (a remnant of the elastic analogy). Because the theory of elasticity is still in the background, I argue that Maxwell had not yet succeeded in founding part B on purely metrological arguments. However the recourse to the elastic theory is now hidden in the background because the mechanic-elastic analogies are now[31] relegated to the role of illustrations to assist the reader's imagination:

> In using such words as electric momentum and electric elasticity in reference to the known phenomena of the induction of currents and the polarization of dielectrics, I wish merely to direct the mind of the reader to mechanical phenomena which will assist him in understanding the electric ones. All such phrases in the present papers are to be considered as illustrative, not as explanatory.

"A Dynamical Theory" ends[32] with a comparison between the self induction coefficient value of a coil, calculated by using Maxwell's new field theory, and that measured by the committee's experiment at King's College, London, 1863: L = 430165 metres (calculated), L = 456748 metres (measured and averaged by the method of least squares).

This is another piece of evidence that, in the composition of his great paper, Maxwell was also inspired by problems raised by his work with the committee.

3.6. *Velocity of Light and Comparison of Forces in 1868*

The experiment intended to measure Weber's factor, announced as part of the Committee's program in the 1863 *Report*, was delayed until 1868. Two short reviews of two methods of conducting the experiment, one by William Thomson and another by Maxwell himself, were published in the *Report of the Association* for 1869. Besides this, Maxwell presented his method in a Memoir to the Royal Society, published[33] in the *Philosophical Transactions*, in June 1868, with the title: "On a Method of Making a Direct Comparison of Electrostatic and Electromagnetic Force: With A Note on the Electromagnetic Theory of Light". The first part of this Memoir dealt with a description of Maxwell's experiment that he cleverly conceived as a balance between an electrostatic attractive force, (the potential difference obtained by connecting two metal disks to two different points of a conducting wire), and a magnetic repulsive force between two coils of the same wire, respectively fixed to the same two disks.

In the balance equation Maxwell expressed the two forces respectively in terms of Coulomb's and of Ampère's laws and he introduced[34] Weber's factor as a conversion factor between the electrostatic units (Coulomb's law) and the electromagnetic units (Ampère's law) [35]; he will abandon this Gaussian metrology in *A Treatise* in favor of the use of one system at one time. Because he measured in electrostatic units the force produced by the potential gradient across the high resistance (28798 OHMS), and because in the electromagnetic system a resistance has the dimensions of a velocity, he measured Weber's factor[36] either in terms of this resistance or in "metres (sic) per second":

$$v = 288\ 000\ 000 \text{ metres per second.}$$

The average is computed on the basis of eleven measurements. According to Maxwell, the probable error is about one-sixth per cent. Maxwell thus measured with his own hands Weber's factor and considered the result a good approximation of Weber and Kohlrausch's value:

$$v = 310\ 740\ 000 \text{ metres per second.}$$

Following the description of the experiment, Part A deals with the same procedure as his former 1865 paper. By the same procedure as in 1865, he derives a propagation velocity for the sole magnetic field:

$$V^2 = + K/4\pi\mu.$$

There is an evident novelty in Maxwell's way of dealing with part B. Here he rewrites the balance equation for the equilibrium of forces, but, this time, in terms of fields. Then he compares it to the former equation for the balance of forces where, as I have shown, Weber's factor v appeared as a conversion of units. The comparison gives: $v^2 = \mu k/4\pi$, i.e., Weber's factor in terms of k and μ. Equating this result with the (1865) expression for V, the conclusion is[37]:

$$\mu V = v .$$

Adding the statement that, in air, µ is "*assumed equal to unity*" (italics in the original):

$$V = v .$$

My argument for the importance of part B in Maxwell's theory is here significantly validated. It is relevant in fact that, in the theoretical part of his 1868 memoir, he presents[38] his deduction as being by itself a proof for the electromagnetic theory of light:

> The statement of the electromagnetic theory of light in my former paper [Maxwell refers to his 1865 "Dynamical Theory... "] was connected with several other electromagnetic investigations, and was therefore not easily understood when taken by itself. I propose, therefore, to state it in what I think the simplest form, deducing it from admitted facts, and showing the connection between the experiments already described [Maxwell refers to his measure of Weber's factor] and those which determine the velocity of light.

Notice that Maxwell ranks his proof to no less than a deduction "from admitted facts", a homage to Newton's unperishable pronouncement.

In comparison with his 1865 approach, this 1868 procedure presents the advantage of avoiding the derivation of Coulomb's law in its second form from elastic argument. The assimilation of Weber's factor to a ratio between electrostatic and electromagnetic forces brings into Maxwell's theory features that are similar to Weber's theory.[39] However, ambiguous metrological considerations are still present in assuming $\mu = 1$, in air and in every system.

One can argue that for these reasons Maxwell abandoned his 1868 approach in *A Treatise*, in favour of a method founded on a complete theory of the absolute systems and a theory of dimensions of electric and magnetic units.

3.7 Maxwell's Final Achievement: Metrology and Theory of Dimension in A Treatise

In his *Treatise*[40] Maxwell modified his method of proving for the equality between the ratio of units and the velocity of the electromagnetic waves (and of light).

The first volume of Maxwell's great work begins with an introductory chapter[41] about the definitions of quantities and of a system of units, although limited at this point solely to mechanical units:

> Every expression of a Quantity consists of two factors or components. One of these is the name of a certain known quantity of the same kind as the quantity to be expressed, which is taken as a standard of reference. The other component is the number of times the standard is to be taken in order to make up the required quantities. The standard quantity is technically called the unit, and the number is called the numerical value of the quantity.

In the following passage the principle of the invariance of physical laws with any change of units and the principle of dimensional homogeneity are clearly stated:[42]

> In framing a mathematical system we suppose the fundamental units of length, time, and mass to be given, and deduce all the derivative units from these by the simplest attainable definitions. The formulae at which we arrive must be such that a person of any nation, by substituting for the different symbols the numerical values of the quantities as measured by his own national units, would arrive at a true result. Hence, in all scientific studies it is of the greatest importance to employ units belonging to a properly defined system, ...This is most conveniently done by ascertaining the *dimensions* of every unit in terms of the three fundamental units...A knowledge of the dimensions of units furnishes a test which ought to be applied to the equations resulting from any lengthened investigation. The dimensions of every term of such an equation, with respect to each of the three fundamental units, must be the same (italics are mine).

The electrostatic (es) and the electromagnetic (em) systems are regarded as the only systems of any scientific value.[43]

Coulomb's laws for quantities of electricity and for the strength of magnetic pole (i.e.,Coulomb's magnetic law), [44] are here the main foundational laws for the two absolute systems. Coulomb's magnetic law is then to be related to the law of interaction between current carrying wires in order to derive em units for currents in the em system. This relation is here obtained by Maxwell through an original field theory of the mutual potential of two closed circuits.[45] The deduction of the dimensions of current intensity in the em system seems to be a crucial point for Maxwell. He affirms that his em system of units corresponds to that used by Weber.[46]

In Chapter ten of Part four, "Dimensions of Electric Units", Maxwell presented a theory of the dimensional relations among the units of two absolute systems "the electrostatic and the electromagnetic system...the only systems of any scientific value".[47]

In an effort at generalisation, Maxwell decided "to begin by stating those relations between the different units which are common to both systems". He listed the following products that have[48] always

A HISTORICAL ROLE FOR DIMENSIONAL ANALYSIS 63

(i.e. in both systems) the dimension of energy:

quantity of electricity x electric potential;

quantity of free magnetism x magnetic potential;

electrokinetic momentum x electric current.

Other products have the dimensions of energy density. Another general property[49] is manifested in a symmetric arrangement of quantities in two lines, in such an order that:

> the quantities in the first line are derived from e [the electric charge] by the same operations as the corresponding quantities in the second line are derived from m [the magnetic charge]. All the relations given above are true whatever system of units we adopt.

The emphasis on the invariance in both systems of certain operations with dimensions are meant to prepare the ground for Maxwell's new approach to the problem of the ratio of units. In fact, this invariance is put immediatly to use in order to deduce the dimensions of the electric charge in the em system, by combining the above dimensional relations *valid for both systems* with a dimensional relation for the magnetic force valid only in the em system. Maxwell also presented a "Table of Dimensions" of the electromagnetic quantities[50] deduced from the above dimensional relations:

Table of Dimensions

	Symbol	Dimension in Electrostatic System	Dimension in Electromagnetic Syst
Quantity of electr.	e	$[[L^{1/2} M^{1/2} T^{-1}]$	$[L^{1/2} M^{1/2}]$
etc.			

This table is followed by a table of the dimensions of the ratios of quantities[51] which "are in certain cases of scientific importance".

	Symbol	Electrostatic Syst.	Electromagnetic Syst.
......	......	......	......
D/E = specific inductive capacity of dielectric	K	[0]	$[T^2/L^2]$
B/H = magnetic inductive capacity	μ	$[T^2/L^2]$	[0]
E/C = resistance of a conductor	R	[T/L]	[L/T]

From the first table:

a) The ratio between the electrostatic unit and the electromagnetic unit for quantity of electricity has the dimension of a velocity.

From the second table:

b) The "specific inductive capacity of a dielectric K " (related to the former k by: $K = 1/k$), has null dimension in the electrostatic system and is assumed equal to one. It has the inverse-square-of-velocity dimension, in the electromagnetic system and is therefore represented by $1/v^2$.

c) The "magnetic inductive capacity" μ has a property[52] reciprocal to that of k, and is therefore represented by $1/v^2$ in the electrostatic system.

Notice that in the above statements what Maxwell actually proved was that $[K] = 1/[v^2]$ and that $[\mu] = 1/[v^2]$ (as above $[\,]$ are symbols for a dimensional relation).

From the following passage one can reasonably argue that he distinguished between a dimensional (and numerical) equality and a physical equality:

If the units of length, mass, and time are the same in the two systems, the *number* of electrostatic units of electricity contained in one electromagnetic unit is *numerically* equal to a certain velocity, the absolute value of which does not depend on the magnitude of the fundamental units employed. This velocity is an important physical quantity, which we shall denote by the symbol v.[53]

This distinction is confirmed in Chapter XIX, "Comparison of the Electrostatic with the Electromagnetic Units", which opens with Paragraph 768, "Determination of the Number of Electrostatic Units of Electricity in one Electromagnetic Unit". In this paragraph Maxwell stated:

It appears from the table of dimensions, Art. 628, that the number of electrostatic units of electricity in one electromagnetic unit varies inversely as the magnitude of the unit of length, and directly as the magnitude of the unit of the time we adopt. If therefore we determine a velocity which is represented numerically by this number, then, even if we adopt new units of length and time, the number representing this velocity will still be the number of electrostatic units of electricity in one electromagnetic unit, according to the new system of measurement.
This velocity, therefore, which indicates the relation between electrostatic and electromagnetic phenomena, is a natural quantity of definite magnitude, and the measurement of this quantity is one of the most important researches in electricity.[54]

Notice that what Maxwell stressed in the initial passages was only the *dimensional equality* between v and a ratio of units: length/time. This equality has the property that the *measure* of v correctly scales with the change of units: "the absolute magnitude" of this quantity is independent from the choice of units.

However, this is just a necessary but not a sufficient condition for the identification of v with the *physical quantity "velocity"*, "a natural quantity of definite magnitude. In fact, Maxwell's next demonstration aims "to shew that the quantity is *really a velocity* "(my italics). Because of the importance of this part of Maxwell's theory, I report the demonstration [55] in its entirety.

To shew that the quantity we are in search of is really a velocity, we may observe that in the case of two parallel currents the attraction experienced by a length a of one of them is, by Art. 686:
$$F = 2\ CC'a/b,$$
where C, C' are the numerical values of the currents in electromagnetic measures, and b the distance between them. If we make b = 2a, then F = CC'.
Now the quantity transmitted by the current C in the time t is Ct in electromagnetic measure, or nCt in electrostatic measure, if n is the number of electrostatic units in one electromagnetic unit.
Let two small conductors be charged with the quantities of electricity transmitted by the two currents in the time t, and placed at a distance r from each other. The repulsion between them will be
$$\frac{CC'\ n^2 t^2}{r^2}\ .$$
Let the distance r be so chosen that this repulsion is equal to the attraction of the currents, then
$$\frac{CC'\ n^2 t^2}{r^2}\ .$$
Hence $\qquad r = nt \qquad ;$

or the distance r must increase with the time t at the rate n. Hence *n is a velocity, the absolute magnitude of which is the same, whatever units we assume (my italics)*.

Maxwell's conceptual experiment can be summarised in the following way: if, at time t, two small conductors are crossed by currents $C = q/t$ and $C' = q'/t'$ respectively, the electrostatic repulsion between tha charges q, q' can be balanced by the Amperian attraction between currents only if they are reciprocally removed at a velocity n: *the ratio of units n is* "really a velocity".

In the following Paragraph 769, Maxwell presented a different argument in order "to obtain a physical conception of this velocity".

Two plane surfaces charged with an electrostatic surface density sigma and sigma' respectively, moving in their own plane with a parallel and concord velocity v and v' respectively, are taken to be equivalent to two electric current sheets.⁵⁶ The density of current through unit breadth of the surface is: $\frac{sv}{\sigma}$ (in es measure), $\frac{1/n\ sv}{\sigma}$ (in em units).

The electrostatic repulsion between the two electrified surfaces is 2πss' for every unit of area of the opposed surfaces.
The electromagnetic attraction between tha two current-sheets is, by Art. 653, 2π uu' for every unit of area, u and u' being the surface-densities of the currents in electromagnetic measure.

But u= $\frac{1}{n}$σv, and u' = $\frac{1}{n}$σ'v', so that the attraction is 2πσσ' $\frac{vv'}{n^2}$.

The ratio of the attraction to the repulsion is equal to that of *vv'* to *n*. Hence, since the attraction and the repulsion are quantities of the same kind, n must be a quantity of the same kind as v, that is a velocity. If we now suppose the velocity of each of the moving planes to be equal to n, the attraction will be equal to the repulsion, and there will be no mechanical action between them. *Hence we may define the ratio of the electric units to be a velocity*, such that two electrified surfaces, moving in the same direction with this velocity, have no mutual action. Since this velocity is about 300000 kilometers per second, it is impossible to make the experiment above described (italics are mine).⁵⁷

In sum, *v* is *physically* a velocity, because an experiment is conceivable implying an operation "such that two electrified surfaces, moving in the same direction with this velocity, have no mutual action".

In the subsequent Article, Maxwell presented a "real" experiment intended to test the hypothesis in the former conceptual experiment that a moving charged sheet behaves like an electric current.⁵⁸

I argue that, in consequence of the two conceptual experiments above, Maxwell believed he was justified to write:

$$[v] = v,$$

i.e. that the proof was reached that the units ratio represented physically, not just dimensionally, a velocity.[59] The *transitional principle* above was thus duly demonstrated.

However, the velocity appearing in both the above conceptual experiments is a convection velocity, i.e., a relative velocity in the motion of charged macroscopic bodies, not the propagation velocity of waves deduced from Maxwell's equations. This is a flaw in Maxwell's metrology which justified the criticism by the opponents of his theory.[60]

In consequence of the above premises, in Chapter XX, entitled "Electromagnetic Theory of Light",[61] Maxwell's task of connecting via the ether constants the ratio of units with Maxwell's wave velocity V was simplified and reduced to purely formal passages[62]:

1) From the D'Alembert equations of the fields, Maxwell deduced the equation which connects the propagation velocity V to the ethereal constants (sometimes named Maxwell's equation):

$$V = \sqrt{k/\mu} \ .$$

2) Starting from this equation, it is proved that in both Systems: $V = v$.

In Maxwell's words:

If the medium is air, and if we adopt the electrostatic system of measurement, $K = 1$ and $\mu = 1/v^2$, so that $V = v$, or the velocity of propagation is numerically equal to the number of electrostatic units of electricity in one electromagnetic unit. If we adopt the electromagnetic system, $K = 1/v^2$ and $\mu = 1$, so that the equation $V = v$ is still true.

In the following pages of the same Chapter, Maxwell presented a short list of the experiments (including his 1868 experiment)[63] to measure v in the form of a ratio of units.

The following table shows that the experiment confirmed the

numerical equality between the ratio measured by Maxwell and Thomson and the most commonly accepted value of the velocity of light:

Velocity of light (mètres per second)	Ratio of electric units (mètres per second)
Fizeau 314 000 000	Weber........310 740 000
Aberration,&c.,and Sun's Parallax 308 000 000	Maxwell.......288 000 000
Foucault 293 360 000	Thomson.......282 000 000

The remarkable numerical difference between Maxwell's and Thomson's V and Weber's c was justified on account of systematic errors in Weber and Kohlraush's instrumentation.

Maxwell commented:

> It is to be hoped that, by further experiments, the relation between the magnitudes of the two quantities may be more accurately determined.
> In the mean time our theory, which asserts that these two quantities are equal, and assigns a physical reason for this equality, is certainly not contradicted by the comparison of these results such as they are.[64]

In support of his conclusions, Maxwell gives special emphasis to the point that the speed of light was measured through experiments that are totally independent from those related to the ratio of units. "Hence the agreement or disagreement of the values of V and v furnishes a test of the electromagnetic theory of light".[65]

3.8. Comments

What was common to Weber and Maxwell and to a large majority of the middle of the century physicists, was a program for measurements of electrical quantities, which was carried on with laborious and meticulous concern and sometimes with exaggerated accuracy.[66] This program was also encouraged by a policy of national rivalries on technical projects for telegraphic and cable transmission.[67]

This program produced important developments in the methods and conception of physics in general.

Maxwell did not include as parts of his field theory Weber's electrodynamics and the convective conception of conduction current. He then made recourse to dimensional considerations limited to his two systems and to Weber-type arguments in order to prove that the ratio v was equal to a physical velocity. In fact, he had first to prove that v was physically not just dimensionally a velocity before giving significance to its approximated numerical equality with the velocity of light.

For this proof, he assumed as a basic assumption for his dimensional approach that his two systems were the only truly scientific systems of Electromagnetism. Maxwell's assumption soon prompted criticism among many German electrical physicists in the last quarter of the century.[68] Moreover, his theory of the equality between v and the velocity of light V was considered with scepticism by Rowland[69] and repeatedly opposed by William Thomson.[70]

In contrast with this criticism of the dimensional approach to V, Maxwell's homogeneity theorems inspired important lines of research. Between 1900 and 1905, Lord Raleigh successfully applied[71] dimensional analysis to problems of mechanics, heat transfer, optics and electromagnetism. In 1914, E. Buckingham generalised[72] Maxwell's theorem of products with fixed dimensions (cf. above, in *A Treatise*) in the so-called "Π theorem", and founded a theory of physically similar systems, through which dimensional analysis became a useful tool in modern theories of physical similitude and of physical models.

Historically, the homogeneity principle represented a mathematical structure which afforded the symbolisation of physics's laws in a form richer than that presented in the former mere numerical and proportionality relations.[73] In fact, due to the homogeneity principle, symbols in the equations represented a number of units.

Elsewhere I have underlined[74] the interconnection between theory, measures and instrumental operations which issued from this new form of representing the physical laws. Quantitative predictions were among its most important consequences.[75] Once more, this function of finding unitary aspects in otherwise different conceptions, and, thus, of pursuing the evolution of theories, was trustedto a technical factor.

Weber's mathematics operated mainly through algebraic passages with a few total-differential equations, thus allowing a more direct connection between the quantities in the equations and the numerical results provided by his instruments. Although Weber's elements of current and electric "masses" are not observable quantities, their easily integrated counterparts, currents in wires and the corresponding reciprocal forces, are directly testable quantities.

Maxwell is a different case. His theory was predominantly expressed by partial-differential equations, i. e., through space-and-time variations of the electric and magnetic quantities whose only observables are represented by their integrated solutions. The predominance in his theory of this differential form of equations has the effect of locating theoretical entities at a higher theoretical level with respect to observables and of pushing these observables a step further from theory, i.e., from the representative theoretical quantities.[76] This can be seen, for example, in his above mentioned method for proving $v = V$, in his *Note on the Electromagnetic Theory of Light* and in his mode of deducing Ampère's law in his *Treatise*.[77]

These features confer to Maxwell's approach a further degree of freedom (so to speak) with respect to the level of observables, increasing the theory's fecundity and predictive power. As regards the problem of a modern understanding of the relationship between Maxwell's and Weber's velocity, I have shownin an earlier study that, in their historical context, the two velocities are numerically and physically different.[78]

A different problem is the modern understanding of Maxwell's statement of the equality between his ratio of unit and his velocity of light V. Modern views on this Maxwellian equality are various. I quote two views which present contrasting positions.

Arnold Sommerfeld[79] believed that the introduction of three fundamental independent mechanical units in a phenomenal area outside mechanics represented a forced limitation of the conventionally free choice of fundamental units and that it was inspired by mechanical preconceptions. Referring to a paradoxical example originally invented by J. Wallot, Sommerfeld proved that even in mechanics, by over determining the system of equations for the definition of units, it was possible to deduce a Maxwellian ratio between different units of mass, a ratio dimensionally equal to a velocity; but this ratio was evidently void of any physical significance! Thus, in Sommerfeld's view, Maxwell's discovery of the equality was a mere chance result, deprived of any basic physical significance.

In contrast with this view, Wolfgang K. H. Panofsky and Melba Phillips illustrate[80] the fact that (*pace* Weber) the number of fundamental independent units and independent dimensions is arbitrary even in classical mechanics, and the specific choice of three units is only suggested by convenience. In the three-unit system, the fact that a constant c, the velocity of light in a vacuum, appears as a fundamental constant of the theory[81] is taken by Panofsky and Phillips as an indication that physical laws "scale" correctly over arbitrary magnitudes only if ratio of length and time are held constant. The authors conclude that "in this property em theory exhibits a feature which special relativity extends to all laws of physics".[82]

Maxwell's approach to the velocity of light through the ratio of two absolute units, proved thus to be fruitful for Einstein's theory of special relativity.

Panofsky and Phillips also prove that, in the so-called mks system, the charge densities ρ_{esu} and current densities J_{emu} obey the relation:

$$J_{emu} = \rho_{esu} u / c, \quad c \text{ velocity of light.}[83]$$

It is worth remarking that if $\rho_{esu} u = J_{esu}$, Maxwell's statement of the equality between a ratio of em and es units for current intensity and the velocity of light is confirmed.

However, it should be said that Maxwell never accepted the microscopic convective conception of the conduction current[84] as expressed in the relation: $\rho_{esu} u = J_{esu}$. Consequently, he could not perform the passage necessary to reach the relation presented by Panofsky and Phillips. As I argued above, this is the reason why he resorted to the dimensional approach and to Weber-type conceptual experiments.

As Buchwald has shown in great detail in his book,[85] Maxwell could not insert in his theory a microscopic convective conception of current, i.e., the conception that a current consists of microscopic charges in motion in a wire.[86] This conception is manifested[87] in Weber's equation: $i = ev/ds$

The similar expression adopted by Maxwell in the above expression, $u = 1/n\sigma v$, concerned the motion of macroscopic charges and represented a conceptual experiment of the Rowland-type, the magnetic effects produced by the motion of a macroscopic charge.

In my view, considered in historical perspective, Weber's and Maxwell's attempts to introduce into electricity the Newtonian scheme of space as a stage for bodies moving in time under the action of a force, and the implicit separation of space from time produced the unexpected and contrasting result of a space-time connection through a constant velocity. It thus proved the fallacy of classical mechanics as a general scheme for the physical world. This fallacy was common to both Weber's and Maxwell's theories in spite of the fact of their difference in their characterising this stage either as an empty space or an ether.

The discovery of c, besides being of fundamental importance for a fundamental law of electrodynamics, was interpreted by Weber as having a metrological role in new formulations of the mechanical units of space and time and of mechanics itself. Weber interpreted the above discovery as indicating a new connection between spacial and temporal units, which allowed a reduction from three to two of the fundamental mechanical quantities and units.

In the context of Maxwell's science, the presence of a velocity numerically equal to Weber's ratio but physically different from it, was interpreted as indicating a propagation velocity of em waves in ether. The Maxwellian interpretation was conditioned by a limitation in the conventionally free choice of absolute systems, somehow partially depriving the meaning of *c* of Weber's more fundamental metrological significance. Though less direct than Weber's, Maxwell's approach had the enormous historical merit of introducing into physics the electromagneitic theory of light.

In his theory of special relativity, it was Einstein who partially reconciled Weber's and Maxwell's approaches, by assuming that Maxwell's constant velocity of light in a vacuum was a fundamental invariant constituent of his theory, which played an important part in describing the mechanical and electrical phenomena involved in the transformation of space-time and of electromagnetic quantities[88] between two reference frames in relative motion.

After purifying both theories of their primitive mechanical and limited metrological features, Einstein thus unified both Weber's metrical and Maxwell's optical interpretations.

CHAPTER 4

PROBLEMS OF THEORETICAL PHYSICS IN THE SECOND HALF OF NINETEENTH CENTURY

4. 1. A Consideration of Maxwell's Innovative Ideas on Physical Theories

Many traditional histories fail to point out that J.C. Maxwell (1831-1879) was innovative not only in his new field theory of electricity and magnetism, but also in his idea of a physical theory.[1]

In his analogical view of theory, Maxwell thought that the search for analogies between mathematical laws in a known area of physics and unknown physical laws was a source of "physical conceptions" because mathematics was therein presented to the mind in an "embodied" form. One of the most fecund of these analogies was William Thomson's parallelism between the Laplacian operator of temperature in heat propagation and the Laplacian operator of electric field in electrostatics: both operators are equal to zero. Since temperature is a field quantity, this analogy suggested to him a field theory of electrostatics. Another example is Maxwell's analogy between an elastic fluid in vortex motion and the electromagnetic field in a vacuum. In this case, hydrodynamic equations for vortex motion furnished a model for Maxwell's celebrated equations.

As to Maxwell's views on the methods of mathematical physics, he warned against the danger that "analytical subtleties" might "draw aside the mind from the subject" and that "mere symbols do not readily adapt themselves to the phenomena to be explained". Not that he discouraged the use of mathematics in physics, but he proposed a new relation between mathematical and physical conceptions.[2] Analogies presented the mathematical laws of phenomena in "an embodied form", a form that he labelled as "embodied mathematics":

> In this outline of Faraday's ideas, as they appear from a mathematical point of view ...my aim has been to present the mathematical ideas to the mind in an embodied form, as systems of lines or surfaces, and not as mere symbols, which neither convey the same ideas nor readily adapt themselves to the phenomenon to be explained.[3]

According to Maxwell, analogies were of various types. Some suggested "temporary theories", i.e., provisional theories which he found very useful while awaiting a "final true theory". He insisted that the scientist could temporarily use "physical analogies" as a helpful method which would suggest provisional ("temporary") theories:

> The chief merit of a temporary theory is that it shall guide experiments, without impeding the progress of a true theory when it appears. [4]

Notice that Maxwell still believed in the final arrival of a true theory. In contrast to the presumed uniqueness of the latter, he conceded that there could be multiple provisional theories in as much as they were ways of looking at a subject:

> it is a good thing to have two ways of looking at a subject, and to admit that there are two ways of looking at it.[5]

Thus, in the context of his analogical conception of theories, Maxwell presented his initial view of theoretical pluralism.[6] He composed his main opus, *A Treatise on Electricity and Magnetism*, in compliance with the above view. Pluralism was the reason for his caution about a theory confirmed by an experiment. Conversely, he could consider an experiment valid independently from its relation to a given theory. This explains why, in his *Treatise*, Maxwell referred to Weber's experiments as perfectly compatible with his own theory, in spite of the fact that Weber considered them favourable to an action-at-distance theory. Evidently, Maxwell thought that the same experiment might support many provisional theories. This point was underlined by Ludwig Boltzmann.[7]

Looked at from the bottom up, theoretical pluralism implied that phenomena do not unambiguously determine a physical theory; let us call this aspect: *the underdeterminateness of the empirical basis* (In the following: UDT). Theoretical pluralism and UDT represented important stages for the future development of a theoretical physics.

In fact, Maxwell's UDT concept was echoed by Hertz and Boltzmann but not by Helmholtz, although Helmholtz was the first to introduce a Maxwell-like theory of electrodynamics in Germany. Let us notice in passing that Maxwell's analogical view of theories represented a counter trend to the methods of mathematical physics mainly to the French methods and an initial foundation for the methods of a "dynamical" theory.

The upsetting novelty of Maxwell's method did not escape Poincaré's attention, and he wanted to warn his French colleagues that they would not find in Maxwell's magnum opus what they expected: "un ensemble théorique logique et précis".

Quite on the contrary:

..le savant anglais ne cherche pas à construire un édifice unique, définif et bien ordonné, il semble plutot qu'il élève un grand nombre de constructions provisoires et indépendants, entre lesquelles les communications sont difficiles et quelquefois impossibles.[8]

In his *Treatise*,[9] Maxwell adopted Lagrange's equations in order "to bring electrical phenomena within the province of dynamics".[10] In order to deduce his equations of the electromagnetic field, he applied Lagrenge's method and its extension in the investigation of Thomson and Tait to the exploration of the space around an electromagnetic system.

As is known, Lagrange introduced generalised co-ordinates in place of the traditional parameters of motion of the Newtonian-Laplacian mechanics. Instead of using quantities representing positions, constraint-forces, velocities, etc., of the particles, in Lagrange's method the motion of the various parts of a system is described "[by] eliminating these quantities from the final equations". This allowed him "to avoid the explicit consideration of the motion of any part of the system, except the co-ordinates or variables on which the motion of the whole depends".[11]

If it is true that, in Lagrange's method, energies must be expressed in terms of some set of generalised co-ordinates and velocities, these quantities, however, need not directly represent an actual mechanical state. Maxwell selected appropriate expressions for the field-energy densities, choosing to represent as kinetic and potential the energies of the magnetic and electric fields respectively.[12] Thus it is appropriate to speak of Maxwell's *dynamical approach* (henceforth DA) to his electromagnetic-field theory. According to Maxwell, to be "dynamical", a theory needed only provide expressions for the kinetic and potential energies for the system but the co-ordinates and velocities of its component parts did not need directly represent an actual mechanical state.[13] He affirmed that in his approach to the electromagnetic-field theory his aim was to cultivate his dynamical ideas, and, for this theory he used the terms "*dynamical theory*" and considered his equations as a special case of "equations of motion of a connected system".[14]

After his first attempt to form a theory of the actual motions and stresses in an elasto-electric ether in his 1860-61 paper "On Physical Lines of Force", it was through his DA that Maxwell achieved his great success in his theory of the electromagnetic field.[15]

This is clearly stated in one of the most important chapters of his *Treatise*:

> What I propose now to do is to examine the consequences of the assumption that the phenomena of the electric current are those of a moving system, the motion being communicated from one part of the system to another by forces, *the nature and laws of which we do not yet even attempt to define, because we can eliminate these forces from the equations of motion by the method given by Lagrange for any connected system.*
>
> In the next five chapters of this treatise I propose to deduce the main structure of the theory of electricity from a dynamical hypothesis of this kind, instead of following the path which has led Weber and other investigators to many remarkable discoveries and experiments, and to conceptions, some of which are as beautiful as they are bold [italics are mine].[16]

That his DA allows the theory to neglect the inner mechanism of the motions which generate the electromagnetic field is clearly stated in the following passage:

> In this outline of the fundamental principles of the dynamics of a connected system, we have kept out of view the mechanism by which the parts of the system are connected. We have not even written down a set of equations to indicate how the motion of any part of the system depends on the variation of the variables. We have confined our attention to the variables, their velocities and momenta, and the forces which act on the I pieces representing the variables. Our only assumptions are that the connections of the system are such that the time is not explicitly contained in the equations of condition, and that the principle of the conservation of energy is applicable to the system.[17]

In chapter IX of his *Treatise*, "General equations of the electromagnetic field", the DA is defined as the method which identifies an electromagnetic system with a Dynamical System:

In our theoretical discussion of electrodynamics we began by assuming that a system of circuits carrying electric currents is a dynamical system, in which the currents may be regarded as velocities, and in which the coordinates corresponding to these velocities do not themselves appear in the equations. It follows from this that the kinetic energy of the system, in so far as it depends on the currents, is a homogeneous quadratic function of the currents, in which the coefficients depend only on the form and relative position of the circuits. Assuming these coefficients to be known, by experiment or otherwise, we deduced, by purely dynamical reasoning, the laws of the induction of currents, and of electromagnetic attraction. In this investigation we introduced the conceptions of the electrokinetic energy of a system of currents, of the electromagnetic momentum of a circuit, and of the mutual potential of two circuits.[18]

It is worth remarking that Maxwell's DA was not compatible with a conception of the particulate nature of electricity. As Jed Buchwald has shown in detail in his book,[19] Maxwell could not find a place in his theory for a microscopic convective conception of the conduction current, i.e., the idea that this current consists of microscopic charges in motion in a wire.[20]

It was through his special usage of Lagrange's equation that Maxwell evaded the burning difficulties of having to describe in detail the motions of ether and of the electric particles, even though he succeeded in establishing a quasi-mechanical theory of electromagnetism.

Although one must acknowledge that Maxwell explicitly considered himself a follower of the atomic constitution of bodies,[21] one must also take into account that in many passages he manifested his conviction that the statistical approach to gas theory was a method congenial to his idea of a dynamical theory.[22]

In order to support this thesis I will refer here to a few passages in Maxwell's kinetic papers. In his 1873 "On the Dynamical Evidence of the Molecular Constitution of Bodies", he returned to his preferred method, "the dynamical explanation of phenomena":

...when a physical phenomenon can be completely described as a change in the configuration and motion of a material system, *the dynamical explanation* of that phenomenon is said to be complete. We cannot conceive any further explanation to be either necessary, desirable, or possible, for as soon as we know what is meant by the words configuration, motion, mass, and force, we see that the ideas which they represent are so elementary that they cannot be explained by means of anything else [italics are mine].[23]

Notice that, in the above passage, Maxwell considered that "the dynamical explanation" was complete in itself because he did not consider any further explanation of the mechanical-atomistic type either possible or desirable.

In an essay entitled "Atom",[24] he justified his criticism of any explanation of this type through the difficulty of explaining by the "small hard body" the variety of spectroscopic lines in gases:

> The small hard body imagined by Lucretius and adopted by Newton, was invented for the express purpose of accounting for the permanence of the properties of bodies. *But it fails to account for the vibrations of a molecule as revealed by the spectroscope.*[25] [italics are mine].

In the continuation of the passage, Maxwell expressed his agreement with Helmholtz's theory of the "vortex ring", and with Thomson's related model of a vortex atom, "imagined as the true form of atom by Thomson [which] satisfies more of the conditions than any atom hitherto imagined".[26]

In his 1870 address to the mathematical and physics Sections of the British Association,[27] after citing Thomson's theory as the one which "seeks the properties of molecules in the ring-vortices of a uniform, frictionless, incompressible fluid", he added:

> If a theory of this kind should be found, after conquering the enormous mathematical difficulties of the subject, to represent in any degree the actual properties of matter, it will stand in a very different scientific position from those theories of molecular action which are formed by investing the molecule with an arbitrary system of central forces invented expressly to account for the observed phenomena....even in the present undeveloped state of the theory, the contemplation of the individuality and indestructibility of a ring-vortex in a perfect fluid cannot fail *to disturb the com-*

monly received opinion that a molecule, in order to be permanent, must be a very hard body [28] (my italics).

I argue that Maxwell's choice of Thomson's vortex theory was motivated by the same ideas expressed in his DA, a view supported by the following passage from the above 1870 address:

> ..the greatest recommendation of this theory {Thomson's theory} from a philosophical point of view, is that its success in explaining phenomena does not depend on the ingenuity with which it contrives "save appearances", by introducing first one hypothetical force and then another. When the vortex atom is once set in motion, all its properties are absolutely fixed and determined by the laws of motion of the primitive fluid, which are fully expressed in the fundamental equations.[29]

The fact that the vortex atom avoided intermolecular forces was for Maxwell another point of merit for the vortex theory, one which accrued to its philosophical value.

In an appendix added on May 1879 to one of his last essays on his kinetic theory,[30] Maxwell clearly described his method in the following passage:

> The method which I have employed throughout is a purely statistical one. It considers the mean values of certain functions of the velocities within a given element of the medium, *but it never attempts to trace the motion of a molecule, not even as far as to estimate the length of its mean path*. Hence all the equations are expressed in the form of a differential calculus, in which the phenomena at a certain place are connected *with the space variations of certain quantities at that place*, but in which *no quantity appears which explicitly involves the conditions of things at a finite distance from that place* (my italics).[31]

In the italicised passage, I want to call attention to Maxwell's pretension not to introduce mechanical features in the molecular motions, not even the mean path length, and to his emphasis on his local approach (implying continuity) to the molecular space. He added that in this work he had adopted the same method[32] used in his fundamental paper "Dynamical Theory of Gases" (1867).

These ideas received an outstanding confirmation in his 1877 booklet "Matter and Motion"[33]. Thereafter Maxwell's criticism of an atomistic approach was extended to the whole of micro-physics: "The investigation of the mode in which the minute particles of bodies act on each other..." was based on the hypothesis that "the configuration, motion, or action of the material systems is of a certain definite kind". But, for him, hypotheses of this type led to uncertain results: even if their experimental confirmation were possible, it would not represent a proof of their correctness, because the possibility could not be excluded that this confirmation was also compatible with different hypotheses.

In Maxwell's own words:

If...the configuration, motion, or action of the material system is of a certain definite kind, and if the results of this hypothesis agree with the phenomena, then, unless we can prove that no other hypothesis would account for the phenomena, we must still admit the possibility of our hypothesis being a wrong one.[34]

In a few words, Maxwell cast a judgement of fundamental ambiguity on the method of framing hypotheses about "the configuration, motion, or action of the material system", for the reason that experiments concerning the proof of hypotheses of this type are not crucial.[35]

He founded his DA on the adoption of a different method. It did not require him to abandon hypotheses altogether, but in framing only those ones which possessed the utmost generality, i. e., those related to "the most general properties of material systems". Among the hypotheses possessing general properties, one is extremely general:

...the extremely general [hypothesis].... is...that the phenomena to be investigated depend on the configuration and motion of a material system... If our hypothesis is the extremely general one that the phenomena to be investigated depend on the configuration and motion of a material system, then if we are able to deduce any available result from such an hypothesis, we may safely apply them to the phenomena before us.[36]

Notice that the extremely general hypothesis of the passage above coincides with the dynamical explanation of a passage (above quoted) from his 1873 paper "On the Dynamical Evidence of the Molecular Constitution of Bodies", one of his main papers on his "Molecular Science", his peculiar form of micro-physics.

In contrast with Maxwell's above difficulty with the conception of atoms and molecules as "small hard bodies", one can also find many quotations which manifest his belief in the molecular constitution of bodies. I suggest that the contrast is only an apparent one because Maxwell did not deny the molecular constitution of bodies, but that he excluded the atomistic approach to theory as a reliable study of such constitution. He favoured instead his DA and the related statistical approach.[37] The point is that, whereas on the ontological level he was a believer in the atomistic constitution of bodies, on the matter of method he discredited an atomistic approach for the above mentioned reasons.

This remark leads us to reflect in passim on the role that Maxwell attributed to the relation between his statistical laws and mechanics. It has been rightly remarked that he highlighted a disjunction between the laws of mechanics and the second law of thermodynamics.[38] The second law of thermodynamics is a statistical law which applies to systems of molecules, not to the fluctuations of individual molecules. He interpreted the statistical method on the phenomenological level by denying that probability can be interpreted as uncertainty of conditions. Thus he contrasted the efficaciousness of his statistical approach with the failure of a mechanistic reductionistic approach to gas theory. Consequently, it can be argued that Maxwell's famous demon was intended as a demonstration that a mechanical reductionistic approach to gas theory led to the paradox of denying the second thermodynamic law. As a higher order justification for this failure he contended that Newton's celebrated second approach to the testing of the force-acceleration law can apply also to microphysics.[39]

4.2. Helmholtz's Secularisation of Kant's A-Priori

The works of Hermann von Helmholtz (1821-1894) represent a pivotal point in the development of both theoretical physics and its related epistemology at the end of the last century.[40]

As is known, he devoted a remarkable part of his work to an analysis of the Kantian problem of the a priori in physics. Kant did not influence Helmholtz through the mediation of the romantic philosophy of Shelling and Schopenhauer, less then ever through the Hegelian *Naturphilosophie*, which Helmholtz vehemently contested.

Through Helmholtz's contributions, physics was reborn from the bosom of physiology.[41] He met Kant right in the middle of his physiological research[42] and originally contributed to a refinement of the role of the a priori in science. His optical research and his theory of vision confirm this statement. In the background, Johannes Müller's theory of the specific energy of sensory nerves inspired Helmholtz's physiological research and his epistemology:

> The stimulation of the optic nerve produces only sensations of light whether that stimulation be caused by objective light (vibrations in the ether), by an electrical current through the eye, by pressure on the eyeball, or by rapid directional changes of the eye.[43]

In 1878, starting from the above premises, although admitting that external influences do affect our senses, Helmholtz denied that the eye, and our senses in general, are passive receptors of a supposedly faithful image of the world:

> Inasmuch as the quality of our sensations gives us a report of what is peculiar to the external influence by which it is excited, it may count as a symbol of it, but not as an image. For from an image one requires some kind of alikeness with the object of which it is an image.[44]

In the case of vision, the only criterion which allows the scientist to somehow relate the subjective "lux" in the eye to the external agent "lumen" is the regularity by which concepts, such as lumen, are related in a theory, and the various sensations of lux (representations) are correspondingly related in the eye. This *parallelism of regularities* (parallelism of laws) holds in general:

> Every law of nature asserts that upon preconditions alike in a certain respect, there always follow consequences which are alike in a certain other respect. Since like things are indicated in our world of sensations by like signs, an equally regular sequence will also correspond in the domain of our sensations to the sequence of like effects by law of nature upon like causes.[45]

Müller's research on the physiology of vision led Helmholtz's interests towards the general epistemological problems of perception and to Kant's doctrine of the a priori, which he labelled as "forms of intuiting and thinking given prior to all experience".

> Johannes Müller's investigations into the physiology of the senses...summarised in his law *of specific energy of sensory nerves* has now brought the fullest confirmation [of Kant's doctrine]; one can almost say to an unexpected degree.[46]

However, according to Helmholtz, this confirmation somehow modified Kant's position, by adding a distinction which was foreign to Kant. It amounted to the resolution into two specifications of the a priori forms of intuition: a *general form* and a *narrower specification*.

According to Helmholtz, what was truly a-priori in every sensory perception is a *general form devoid of any content*, as exemplified in the spatial perception of place or in the visual perception of an aggregate of coloured surfaces, which always compose our visual field.

Quite differently, the *narrower specification* concerned the various types of spaces described by the axioms of geometry or, in the visual perception, its content itself, "the particular colours which appear on this or that occasion, their arrangement or sequence".

In an attempt to better explain how the resolution above operates, he thus exemplified its inhering in the spatial and in the optical perceptions:

> Everything our eye sees, is an aggregate of coloured surfaces in the visual field-that is our [general] form of intuition. The particular colours which appear on this or that occasion their arrangement and sequence this is the result of external influences and is not determined by any law of our makeup.
> Similarly, from the fact that space is a form of intuiting, nothing whatever follows about the facts expressed by the axioms. If such propositions are taken to be non empirical ones, but to belong instead to the necessary form of intuition, then this is a further particular [narrower] specification of the general form of space; and *those grounds which allowed the conclusion that the form of intuition of space is transcendental, do not necessarily for that reason already suffice to prove, at the same time, that the axioms too are of transcendental origin* [italics are mine].[47]

For Helmholtz, although the general form of spacial perception is truly a priori, this does not entail that the axioms of a specific geometry are also given a-priori. On the contrary, the specification that Kant introduced into our spatial intuition (i.e., its three-dimensionally) "limits the form of intuition of space in such a way that it can no longer absorb every thinkable concept, if geometry is at all supposed to be applicable to the actual world...".

The axioms of Euclidean geometry represent the intuition's narrower specification and thus the three-dimensionality of space represents its empirical specification.[48] In short, a form of intuition which limits our perception of space to its three-dimensional formulation is too narrow (too full of a particular content) to represent all possible contents of our experience and, as such, cannot be a truly a-priori form:

These forms must be devoid of content and free to an extent sufficient for adsorbing any content whatsoever that can enter the relevant form of perception.[49]

Applying to the study of perceptions in general his distinction of the Kantian a-priori forms in spatial perceptions, Helmholtz found that, between the various kinds of sensations, there exist differences in the form of two neatly distinguished types: one form concerned sensations belonging to different senses, such as sight and sound; he called this a *difference in the modality (of sensations)*. The other form distinguished sensations of the same type, and he called it a *difference in quality*.

The *modality* form belonged to the general form of intuitions, the true Kantian a-priori. For instance, optical modalities are distinguished from the acoustical ones by their characteristic features.

One can argue that, in Helmholtz's conception, the known eye's lack of distinction between pure and mixed colours could not be considered any more as the eye's imperfection or incapability (as it was for Newton) because this lack of distinction was rather the eye's own modality, which differentiated vision from other senses, e. g., from hearing.

Distinctions in modalities being inaccessible to linguistic distinctions, such processes were accessible to science only via an operational approach. In this approach to the study of perceptions, as distinguished from a merely linguistic approach, Helmholtz opened a new field of inquiry which still bears fruit in modern research.

Helmholtz considered his contribution, the result of his own physiological investigations, an important advance on Kant because "the processes related to the modality features of perceptions had to remain still unformulable in words, and unknown and inaccessible to philosophy, as long as [Kant] investigated only cognitions finding their expression in language".[50]

He was aware that he had advanced Kant on his own ground: "here Kant was not critical enough in his critique".[51]

The distinction between differences in modality and quality represented Helmholtz's undeniable merit. We are here presented with the remarkable historical case of a philosophical advance achieved through a technical approach to the problem of perceptions.

In his rectorial address at the Berlin University, *The Facts in Perception,* Helmholtz explicitly stated that the same philosophical problem was common to science and philosophy, although the two disciplines encountered it proceeding from two opposite directions:

What is true in our intuition and thought? In what sense do our representations correspond to actuality? Philosophy and natural science encounter this problem from two opposite sides, it is a task common to both.[52]

He concluded that his views were fully in accordance with Goethe's position:

I consider it a propitious sign that Goethe, both here and in other matters, finds himself with me on the same path.. His theory of colour can be considered *an attempt to save the immediate truth of sensory impressions* from attacks by science[(italics are mine].[53]

At the close of his aectorial Address, he emphasised that his favourite approach had no less than Goethe and Kirchhoff as its supporters. He agreed with the poet that one should simply demand from science that "it should be only an artistic arrangement of the facts and form no abstract concepts going beyond this". He deviated from Goethe's charge that concepts obscure facts, limiting this charge to those abstract concepts which do not possess a corresponding perception:

Concerning the accuse of obscuring, it takes place when we deal with abstract concepts and we don't understand their factual significance, that is, we don't clearly realise what new observable relation between phenomena those concepts express, i.e., which law derives from them.

He also confirmed Kirchhoff's statement that "the task of the most abstract amongst the natural sciences, namely mechanics [is] to describe completely and in the simplest manner the motions occurring in nature".[54]

In his research he always studied the correlation between the presumed external agent and the observable effects on our sensation, with the aim of describing these effects. An ideal description should thus achieve a perfect adequacy between concepts and the corresponding perceptions.

There is some agreement among scholars on the point[55] that Helmholtz advanced a physiological and psychological interpretation of those form of intuition which Kant had credited to a transcendental aspect of knowledge. Limiting the theory's scope merely to a description of phenomena, Helmholtz somehow diminished the cognitive import of the Kantian transcendental forms of intuition and of the Kantian *Kategorien*, thus contributing to their, so-to-speak, "secularisation".

Although Helmholtz maintained that - pace Goethe - his philosophy advanced a complete synthesis of concepts and perceptions, the doubt remained that concepts are grasped only by the intellect, while perceptions are given to us in their intuitive immediacy.

Thus, in my opinion, Helmholtz's thought stands on the watershed between two great traditions in the development of modern physics. On the one hand, through the attention given to perceptions per se and to psychology, he opened the way to Mach's phenomenology. On the other hand, with his special defence of Kant's apriorism, Helmholtz favoured Hertz's philosophy and Hertz's re-evaluation of Kant's a-priori. Mediating between the two tradition, Boltzmann introduced a hereditary Darwinian interpretation of the a-priori in physics.[56]

4.3. Theory and Experiment in Hertz's Holistic Conception of Theoretical Physics

In the introductory pages of his last work, *The Principles of Mechanics Presented in a New Form,* Heinrich Hertz (1857-1894) presented his interested readers with a view of mechanics that took account of the problems of nineteenth-century mechanistic conceptions and, at the same time, opened the way to the epistemological debate of the turn of the century:

> All physicists agree that the problem of physics consists in tracing the phenomena of nature to the simple laws of mechanics. But there is not the same agreement as to what these simple laws are... it is just here that we no longer find any general agreement. Hence there arise actual differences of opinion as to whether this or that assumption is in accordance with the usual system of mechanics, or not. It is in the treatment of new problems that we find a real bar to progress.[57]

Although a mechanical reductionism is still in Hertz's program, his experiments on the electromagnetic waves impressed him with a new theory-experiment relation, i. e., with the exigency of devising a new conception for physical theory.

In Hertz's view, because of the doubts that recent developments had cast on old ontologies, a foundation for a new conception of physics should have been based only on predictive features, "the anticipation of future events". Because Helmholtz had viewed the parallelism of laws as the unique condition for this anticipation, Hertz paraphrases this parallelism in some detail:

> In endeavouring thus to draw inferences as to the future from the past, we always adopt the following process. We form for ourselves images or symbols of external objects; and the form which we give them is such that the necessary consequences of the images in thought are always the images of the necessary consequences in nature of the things pictured. In order that this requirement may be satisfied, there must be a certain conformity between nature and our thought. Experience teaches us that the requirement can be satisfied, and hence that such a conformity does in fact exist.[58]

However, according to Hertz, Helmholtz' above mentioned requirement does not unambiguously determine the choice of an appropriate theory. A *multiplicity of representations* is thus consistent with the Helmholtian requirement:

> The images [*Bilder*] which we form of things are not determined without ambiguity by the requirement that the consequences of the images must be the images of the consequences. Various images of the same objects are possible, and these images may differ in various aspects.[59]

A complex of perceptions does not univocally determine the choice of a theory, for there exists a multiplicity of representations consistent with a term-to-term correspondence between concepts in a theory and perceptions. In other words, the above Helmholtian requirement of a strict parallelism between concepts and perceptions does not hold.

According to Hertz, Helmholtz's parallelism is not only underdetermined, but also impossible in general, if theory is to be limited to observable quantities. Hidden entities, such as electromagnetic ether, have for Hertz a fundamental role:

> If we try to understand the motion of bodies around us and to refer them to simple and clear rules, paying attention only to what can be *directly observed*, our attempt [to construct an appropriate theory] will in general fail. We soon became aware that *the totality of things visible and tangible* do not form an universe conformable to law, in which the same results always follow from the same conditions. We become convinced that the manifold of the actual universe must be greater than the manifold of the universe which is directly revealed to us by our senses [italics are mine].[60]

Only the introduction of hidden quantities allows parallelism to reach the status of a general principle. The construction of a general theoretical system encompassing mechanics and electrodynamics, Hertz's final program in the *Prinzipien*, implied the concept of a *hidden substance* as its unifying element.

In consequence of the above, theories are undetermined or underdetermined with respect to their empirical referents. Underdeterminationism (UDT), the impossibility of deciding about a valid theory only on the basis of empirical considerations, represented an important element in the distinction between Helmholtz's and Hertz's epistemologies; for Hertz takes his start from UDT as the basis for a further development of internal (non-empirical) criteria of validation as indispensable requirements in theoretical physics. These requirements function as new selection criteria for the choice of an adequate theory and are then to be considered higher level formal criteria, consistent with Hertz's ideal of an adequate theory.

Hertz discussed in detail these requirements in his introduction to the *Prinzipien* . In order to be able to make predictions, we form images (Bilder) of things, but we form them in accordance with a given set of requirements. At least one of these, *permissibility* (Zuläs-

sigkeit) is a-priori in Kant's sense:

What enters into the images, in order that they may be permissible (Zulässig), is given by the nature of our mind; we should at once denote as inadmissible all images which implicitly contradict the laws of our thought.[61]

Zulässigkeit is acceptable in as much as theories are *Bilder*, i.e. they concern "images of our own creation not nature". Simplicity is one and the most relevant feature of the *Zulässigkeit* requirement:

It is true we cannot a priori demand from nature simplicity, nor can we judge what in her opinion is simple. But with regard to images of our own creation we can lay down requirements. We are justified in deciding that if our images are well adapted to the things, the actual relations of the things must be represented by simple relations between the images.[62]

In his prefatory note to book one of *Prinzipien*, Hertz showed how, by shaping his theory through the requirements above, he was inspired by the Kantian conception of a-priori judgements:

The subject matter of the first book is completely independent of experience. *All the assertions made are a-priori judgements in Kant's sense.* They are based upon the laws of internal intuition of, and upon the logical forms followed by, the person who makes the assertions; with his external experience they have no other connection than these intuitions and forms may have[my italics].[63]

He continued, in an orthodox Kantian fashion:

The time of the first book [of *Prinzipien*] is the time of our internal intuition... in itself it is always an independent variable. The space of the first book is the space as we conceive it. It is therefore the space of Euclid's geometry, with all the properties which this geometry ascribes it [my italics].[64]

In his acceptance of an a-priori Euclidean space and an intuitive time, Hertz neglected or ignored Helmholtz's criticism of the Kantian a priori. He improved on Helmholtz's notion of the parallelism of laws, but fell back on an a-priori too full of content, the target of his master's criticism.

Given the above mentioned role of internal (non-empirical) constraints, theory assumes the features of a set of axioms, which, rather than emanating from "observables", are formulated prior to them, in the sense of being "constitutive of sense" for the observables themselves.

On this important feature of theory, Hertz is very definite. As I show in the following pages, his experiments became significant when he adopted the *constitutive principle of independent existence* of forces in a vacuum.

Independent existence amounted, in essence, to a renunciation of the classical concept of force as that which mediates at a distance the interaction of bodies.[65] Hertz consistently adopted this position, renouncing forces in his papers on electrodynamics[66] and in his final masterpiece, the *Prinzipien*. In my view, *independent existence* is the main pillar of the Hertzian logic of discovery and, at the same time, the truly innovative content of his epistemology.[67] He took it as a principle of the new electrodynamics, consistent with his assumption of a contiguous action.

In as much as waves of electric force could also be explained by Helmholtz's theory of a dielectric polarisation acting at close range, experience alone was unable to suggest and support Hertz's conception of contiguous action. A striking confirmation thereof is the fact that, more than ten years after Hertz's celebrated experiment, Boltzmann still considered contiguous action as "completely beyond the facts". According to him, "contiguous action... however a-priori it may seem to some, still goes completely beyond the facts and to date remains well beyond what can be elaborated in detail".[68]

Boltzmann was right because no crude fact could directly prove contiguous action. Hertz would have contended that only an a-priori assumption, in the form of a principle, could give support for contiguous action to be shown by an experiment. However, once accepted, this principle revealed a surprising power in directing the experiment towards Hertz's goals.[69] Poincaré, too, doubted that Hertz's experiment was really crucial to Maxwell's theory.

From his great experiment and from his reflections on the *Principles of Mechanics*, Hertz derived a holistic conception of the concepts-facts co-ordination. In consequence of the fact that concepts are logically correlated into the theory and facts are systematically deduced from concepts, *no fact in isolation can be conceived and no experiment does concern an isolated fact. The concept-fact co-ordination concerns the theory as a whole.* All this implied, according to Hertz, that no experiment, as a collection of isolated facts, can be crucial for a single theory, but experiments indicate which, among many theories, is the correct one.

Let us explore in some more detail how Hertz, one of the most expert experimentalists of the last century, conceived of this co-ordination. In his 1889 theoretical paper, "The Forces of Electric Oscillations Treated According to Maxwell's Theory", he presented Maxwell's equations in ether as the core of this theory and he clearly argued for the impossibility of "deducing them from the experiments",[70] thus rejecting inductionism in empirical sciences.

One is thus convinced that only his philosophical ideas could have supported the burden of that engaging shift from traditional action-at-a-distance mechanics to the field conception, a different *type* of theory[71] that Hertz developed in his 1888 research.[71]

No sort of empirical necessity whatever *imposed* this shift; in fact, neither Helmholtz nor Boltzmann (the latter closer on this point to Helmholtz than to Hertz) succeeded in performing the same revolutionary paradigmatic change that Hertz did, although they shared the observational evidence of Hertz's experiments.[72] Hertz himself admitted that other types of theory such as Helmholtz's could almost as well explain his experiment.[73]

I have up to now illustrated Hertz's ideas on the leading role of theory in physical research. However, one should also acknowledge that Hertz's enhancement of the role of theory was no vindication at all of the view that theory alone may constitute an advancement in physics. In his introduction to his experiments, Hertz was very definite on this point : no theory could have foreseen the behaviour of those electric sparks, which allowed him to detect exceptionally rapid electric oscillations and waves:

> Nor, indeed, do I believe that it would have been possible to arrive at any knowledge of these phenomena by the aid of theory alone. For their appearance upon the scene of our experiments depends not only upon their theoretical possibility (German: "theoretischen Möglichkeit"), but also upon a special and surprising property of the electric spark which could not be foreseen by any theory.[74]

One must conclude that a sort of autonomy exists at the empirical level which represents the counterpart of the enhancement of the theory's role. In his theoretical paper published in the *Göttinger Nachrichtungen* in 19 March 1890, he affirmed that experiments are probative on their own, independent of theory:

> What we here indicate as having been accomplished by the experiments is accomplished independently of the correctness of particular theories.

However, he added:

> Nevertheless, there is an obvious connection between the experiments and the theory in connection with which they were really undertaken.[75]

Let us try to untie this epistemological knot in Hertz's view of the theory-experiment relationship; we will discover that Hertz supports a holistic conception of the theory-experiment relationship. In his view, the concepts-facts relationship holds in a direction opposite to the presumed inductive process. In fact, he showed "in what manner the facts which are directly observed can be *systematically* deduced from the formulae; and, hence, by what experience the correctness of the system can be proved"[76] (my italics). Although the physical interpretations of the equations are "facts derived from experience, and experience must be regarded as their proof", the equations' symbols are not related to experience term to term, as if this last consisted of isolated facts:

> It is true, meanwhile, that each separate formula cannot be specially tested by experience, but only the system as a whole. But practically the same holds good for the system of equations of ordinary dynamics. [77]

In his 1888 paper, "On Electromagnetic Waves in Air and their Reflections", he commented on the relation between his experiments and Maxwell's theory:

> I have described the present set of experiments, as also the first set on the propagation of induction, without paying special regard to any particular theory; and indeed, the demonstrative power of the experiment is independent of any particular theory. Nevertheless, it is clear that the experiments amount to so many reasons in favour of that theory of the electromagnetic phenomena which was first developed by Maxwell from Faraday's views.[78]

One is confronted here with the somehow ambiguous statement that experiments have a certain type of autonomy in the face of theory (the demonstrative power, etc.), and yet, they can speak in favour of one or another of the competing theories. In the second part of the statement above, Hertz admits that experiments show something independent from any theory, they show an *opacity* to use a fashionable term to any theory. On the other hand, as a consequence of the holistic conception, theory has an autonomy of its own with respect to the empirical level.

I think that there is no ambiguity if one argues that Hertz meant to highlight a partial dichotomy between theory and experiment. Since the choice among a plurality of theories is not determined by observation, no experiment can be crucial for either one or the other among a plurality of theories; but, even if no experiment can be crucial for a single theory, experiments as a whole do indicate which, among the best theories, is the correct one. Hertz thus denies the conception of cruciality in physics. In this sense we understand Hertz's concluding remark in his 1889 paper: "Maxwell's theory has been found to account most satisfactorily for the majority of the phenomena".[79]

It is then conceivable how the above mentioned reciprocal partial autonomy between experiment and theory is to be understood.[80] In the old view of the theory-experiment relationship, one expected that the experiment was to check and sometimes to falsify a theory's single law. But, if one accepts the holistic conception of theory as a logically connected whole, then the view above is no longer valid. In the UDT conception, even a whole set of disproving experiments cannot, in principle, falsify a theory.

I consider Hertz's ideas an important end-of-century achievement not only in physics but in the conception of empirical science itself[81]. A new relation is established between theory and experiment, which might be considered as one of the characteristic features of the new theoretical physics.

4.4. Mach's Descriptive Ideal and the Elimination of Models

Ernst Mach believed that the physical conditions of a phenomenon can be separated from the physiological conditions through the experiment:

> The sum total of the occurences observable in common by all people with a normal sight we shall call physical optical data (Tatsachen)....The chalk and the flame which previously appeared white also, however, appear yellow when we take a dose of santonin. In this case we regard the yellow which we see, but others do not see, as determined physiologically by a condition appertaining to the more limited sensory complex of the body. Thus the same elements which we observe white, yellow, red etc., are according to the circumstances, sometimes physical, sometimes physiological, [i.e.] sometimes features of bodies and of their behaviour towards other bodies, and sometimes sensations.[82]

The yellow seen by the person who took a dose of santonin cannot be seen by all other persons who, instead, see the chalk either white or yellow, depending on the white or yellow sodium flame by which it is illuminated. It is thus possible to distinguish the first physiological from the second physical occurrence. Goethe and Schopenhauer, according to Mach, made a mistake when they did not sufficiently realise the above distinction.

When Helmholtz asserted that there are "statements which we can make independently of the particular nature of our eyes", he was almost on the same ground as Ernst Mach. In fact, Helmholtz continued, the above is true "when we assert that the lights reflected from cinnabar have a certain wavelength,... with this statement it is...only a matter of relations between the substance and the various ether-wave systems".[83]

There is a however a relevant detail which distinguishes Helmholtz's from Mach's position. In contrast with Helmholtz, Mach would have not considered the ether-waves of Helmholtz's example above as "bodies" of the same type as those "observable by people with a normal sight (Helmholtz)". Ether-waves are generated, according to Mach, through sensation complexes, hence they cannot generate sensations:

It is not bodies which generate sensations, but it is sensations complexes instead which form bodies.[84]

For Mach, Helmholtz's ether-waves are not sensations; rather, they are to be regarded as objects (bodies) formed by sensation complexes, then a *post* as regards sensations, a model which belongs to a different level (layer) with respect to sensations.

Mach is very definite on this matter: ether-waves cannot be considered in turn as the cause of sensations and as their effects, one has to decide. For him, ether-waves are formed out of sensation complexes, and, as such, they cannot be considered as external bodies, the cause of sensations. On this point, as Schlick aptly remarked, Helmholtz seemed to be wavering:

The "external causes" of which Helmholtz ... considered sensations to be effects, are at any rate not these bodies but can instead only be understood to be transcendental things.[85]

One can elucidate the same problem from a slightly different angle. Ether-waves are concepts belonging to theory, and, as regards theory's usefulness, Mach assumed a very cautious attitude. He examined both advantages and dangers in the adoption of a theory:

A theory puts in the place of a fact A in thought always a different but simpler and more familiar fact B, but for the very reasons that it is different, in other relations cannot represent it...... On the other hand, if the agreement of the fact with the idea theoretically representing it extends further, than its inventor originally anticipated, then we may be led by it to unexpected discoveries, of which Hertz's waves offer ready examples, in contrast to the illustrations given above.[86]

Theory has for Mach an auxiliary function, it is a guide in the broadening of an enquiry, for whose validity we have no internal (to theory) hint other than its success. Its validity is a factual one, which we must trust, but cannot prove. In this connection one has to properly situate Mach's often misinterpreted sentence:

> Science completes in thought facts that are only partly given.[87]

Concerning the sense in which this completion operates one has to refer back to Mach's words:

> This is rendered possible by description, for description presupposes the interdependence of the descriptive element: otherwise nothing would be described.[88]

Theories help in achieving the above interdependence, but once this task has been accomplished, their function is exhausted and they are to be eliminated:

> It would appear...not only advisable, but even necessary, with all due recognition to the helpfulness of theoretic ideas in research, yet gradually as the new facts grow familiar, to substitute for indirect description *direct* description, which contains nothing that is unessential and restricts itself absolutely to the abstract apprehension of fact.[89]

One can wonder why theories are necessary if they are destined to be eliminated, thus accomplishing, so to speak, the role of bloody sacrifices in the ancient Greek myth of Ifigenia. Mach explains this ambiguous necessity:

> We must admit that it is not in our power to describe directly every fact on the moment. Indeed we should succumb in utter despair if the whole wealth of facts which we come step by step to know, were presented to us all at once. Happily, only detached and unusual features first strike us, and such we bring nearer to ourselves by *comparison* with every day events[italics in text].[90]

Theories, mainly because of their similarity features, perform a regulative role in the completion of description. For Mach, the efficacy of using a theory is not much connected with the advantages of a "mental visualisation", but rather with its affording "impulses to some accurately determined, often complicated, critical, comparative, or constructive activity, the usual sense-perceptive result of which is a term or member of the concept's scope.... The concept is to the physicist what a musical note is to a piano player....Long and thoroughly practised actions, which have their origins in the necessity of comparing and representing facts by other facts, are thus the very kernel of concepts".[91]

Here Mach has learned a lesson from Helmholtz's research into the role of the body's cinestesis for the analysis of sensations.

Since theories are made up of concepts, Mach asks what concepts are and what is the scope of introducing very abstract concepts, such as Ether:

A wide reaching *abstract concept* is...indispensable in the description of broad fields of facts.

Abstraction has for Mach the merit of generality, a broader coverage of particular cases, which becomes very useful in description.

Mach's reaction to Hertz sheds further light on the meaning of the Machian description. Hertz's statement that the essence of physics is to *prophesise, to predict future occurrences,* is considered by Mach a "too narrow" definition:

The geologist and the palaeontologist, at times the astronomer, and always the historian and the philologist, prophesize, so to speak, *backwards*...Let us say rather: science *completes in thought facts that are only partly given*. This is rendered possible by description, for description presupposes the interdependence of the descriptive elements: otherwise nothing would be described [italics in text]. [92]

Mach's ideal of theories as *complete descriptions* is more general and comprehensive than the above Hertzian definition; besides "prediction" it includes "post-diction". A complete description is "pre-dictive" and "post-dictive" in the sense that whatever can be said about past and future is, as yet, contained in it (description presents an eternal present). The general system of concepts through which prediction is rendered possible, i.e., theory, is correspondingly devoid of any cognitive value per se.[93] Description is for Mach complete and unambiguous.

In the past, science has suffered from incompleteness in description, at times recurring as a remedy for metaphysical ideas (such as Newton's absolute space) in an attempt to overcome its defects. I dare say that this might be Mach's rationale in his historical reconstruction of the development of mechanics.

As Einstein once said, Machian phenomenology has a charm of its own due to its basic scepticism towards any preconceived judgement on the nature and role of science. According to Cassirer,[94] phenomenology also has the merits of helping to supplant naive nineteenth-century realism and of encouraging physicists towards philosophical speculation.

On the other hand, it should be also conceded that the Machian ideal of a complete description which denied any role to the systematic arrangement of concepts beside that of a short-hand transcription, plaid down the fecundity of conceptual innovation, i.e., one of the main features of the theoretical physics. The success of Einsteinian relativity is a striking example thereof.

PART TWO

Electromagnetic Waves

FOREWORD TO PART TWO

This second part presents a review of the delopment of German electrodynamics in the 1870' and a critical evaluation of Hertz's celebrated experiments.

The core and the characteristic features of Hertz's research in electrodynamics was his discovery in 1888 of the propagation *in air* of electric and magnetic force, interpreted from the stand-point of a Maxwell-type theory of contiguous action.[1] Waves of currens in wires were theoretically and experimentally known since Weber, Kirchhoff, von Bezold, Kelvin. Waves in material dielectrics were predicted by Helmholtz's action-at-a-distance theory. Hertz bridged Maxwell's waves with the tradition of current-waves and dielectric-polarisation-waves.

Hertz's conception in 1884 of the unity of electric forces is the origin of his later conversion to the theory of a unique type of force in radiation (see below chapter seven). He thought that such a force was consistent only with Maxwell's theory, which he contrasted with Helmholtz' two-force theory.

I find adequate evidence for my above theses in Hertz's original contribution to electrodynamics and mechanics. An important piece of evidence is presented by Planck's authoritative comment that Hertz's 1884 study "weighed considerably in Hertz's mind in favour of Maxwell's theory ".[2] Elsewhere Plank adds that this study represents a first-rate piece of theoretical work, as impressive in its way as Hertz's later experimental work which eclipsed it, and he regrets that the 1884 work has received so little attention.[3]

In Chapter 8, I analyse Hert's conception of mechanics and I place a special enphasis on his *Bild*-conception of theory considered as an important philosophical qualification of the nature of theoretical physics.

In the Chapter on Boltzmann, I analyse in some detail how Boltzmann modified Hertz's *Bild*-conception of physical theory in order to adapt it to his own philosophical and physical conceptions. Following these conceptions, he constructed his mechanics and gas theory. It is my hope that this analysis might be useful for understanding some controversial aspects of Boltzmann's contributions to physics.

CHAPTER 5

GERMAN ELECTRODYNAMICS IN THE 1870'S

5.1 Circuital Theories and Faraday's Induction

The most successful developments in mathematical physics in Germany were the various derivations from fundamental principles of Michael Faraday's electromagnetic induction law of 1831.

Franz Neumann in 1845 and 1848 derived the electromotive force of induction from the potential of those ponderomotive forces which, according to André-Marie Ampère's theory, acted between a closed circuit and a magnet. Neumann's work was a generalisation of Ampère's theory and a brilliant confirmation of the fecundity of the potential method of Gauss and Green.

Wilhelm Weber took a different approach to the law of induction (See above chapter one, especially sectios 1.4 and 1.5). In 1846 he published his elementary law of action between charged particles that comprehended static, ponderomotive, and inductive phenomena. His program consisted in deriving Ampere's ponderomotive forces between currents from more fundamental hypotheses concerning the elementary forces exerted by positive and negative charge carriers constituting electric currents. To yield Ampere's ponderomotive forces between stationary currents, the elementary law of force had to depend on the components of the relative velocity of two charged particles along their radius vector. The law accounted for the induction of a current when a circuit moved relative to another circuit or magnet. Weber's conception of induction as an effect of an alteration through relative motion of the elementary forces between particles implied that the forces acting on electric particles producing currents were of the same nature as the electrostatic forces between macroscopic charged bodies. In fact, Rudolf Kohlrausch had announced the identity of electrostatic and electromotive forces for steady currents in 1849, an identity that he considered an indication of the reality of the conception of current as the motion of an electric substance.[1]

According to Weber and Kohlrausch, positive and negative electric particles moved in a wire in opposite directions with a maximum relative speed, whose value was measured by C_W, a constant in the law. C_W was close to the velocity of light in a vacuum, c; in fact, $C_W = c\sqrt{2}$. However, because of the conceptual context in which Weber was working, he failed to see in this near numerical coincidence any hint of a hidden relationship between electrical and optical phenomena. Nor did he see one in his conclusion of 1854 in which he stated that his elementary law entailed periodic oscillations of electric currents with a propagation velocity equal to that of light in perfectly conducting circuits. Kirchhoff, too, in 1857 calculated the propagation of current for the case of short wires of small resistance, using expressions for local current and charge gradients. He noticed that the propagation velocity was equal to that of light, but the case of short wires of small resistance was one to which he attached little importance; he was primarily interested in the opposite case of long wires suitable for telegraphy, for which he found no propagation effect.

The case of long wires had been treated a few years earlier, in 1854, in a correspondence between two British physicists, George Gabriel Stokes and William Thomson.[2] From the idea that self-inductance L and capacitance C in the variable regime of currents produce electromotive transient forces, Thomson derived a telegraphic equation. However, the extreme cases that he treated were an aerial wire and a submarine cable, in which either L or C predominated and in which an ohmic resistance is also present. In these conditions no definite velocity of transmission is to be expected for ordinary signals. In the general case, which Thomson did not treat, the velocity, for a small resistance, assumes the well-known value

$v = 1/\sqrt{LC} = c$, the velocity of light in air.

Physicists developed other Weber-type laws, such as the one that Bernhard Riemann formulated in a course of lectures at Göttingen in 1861. Weber and his followers pursued the theory of electricity and magnetism in Ampere's spirit towards the goal of a central force physics. Awakening mathematical physicists to the possibilities latent in the theory of electricity,[3] these laws represented a great advance in electrodynamics and in physics.

In his 1847 memoir on the conservation of force, Hermann Helmholtz gave a different derivation of the same law. There he applied the conservation principle to the interaction of magnets and currents:

When a magnet is moving under the influence of a current, the living force that is gained by the magnet must be communicated by the forces of the potential [*Spannkraften*] that are utilised by the current.[4]

This statement allowed him to express the current J induced in a circuit of resistance W in the form:

$$\int J \, dt = \frac{1}{a} \frac{V_1 - V_2}{W} \quad ,$$

V_1 is the initial potential of the magnet relative to the conductor when the latter carries unit current, V_2 the potential at the end of the displacement, a the mechanical equivalent of heat, and t the time.

When Helmholtz formulated his law he did not know of the potential energy accumulated in the magnetic field of a current, though this incompleteness does not qualify the validity of the law in this special case.[5]

At first Helmholtz considered forces depending on velocity and acceleration to be in contradiction with the conservation of force or energy. He strongly insisted on the incompatibility of Weber's law with general principles.

Since the protracted debate between Helmholtz and Weber had consequences for Hertz's approach to electrodynamics, I will give a short summary of it here.[6]

In 1870 Helmholtz criticised Weber's law on the grounds that it produced an unstable equilibrium in the electricity inside a conductor. Weber replied that this instability would not occur if the relative velocity of the electric particles remained lower than the velocity of light, and that, in any case, the instability would occur only for molecular distances. Helmholtz rejoined in 1873 that by Weber's law a partially charged particle endowed with mass would suffer a delay when moving in the direction of the force; Weber in 1875 rejected the validity of this conclusion. Eventually, in 1881, Helmholtz found a case in which the application of Weber's law yielded an imaginary velocity.

Other German physicists participated in this debate: in 1875 Carl Neumann sided with Weber, in 1875 and in 1876 Rudolf Clausius joined the other side. The strength of Weber's theory can be appreciated by noting that as late as 1884, E. Hoppe still considered it the only true explanation of the presence of an induced current. In light of Helmholtz's many theoretical efforts at dismantling Weber's theory, it is worth mentioning here that in 1872 Helmholtz commended Maxwell's theory on the grounds that it avoided the "anomalous kind of forces which depend not only on the position of the masses but on their motion".[7] When in 1876 Henry Augustus Rowland's experiment pointed to a velocity-dependent force (as an effect of a convection current), Helmholtz put forward the alternative view that the polarisation current produced by motion at a point in the surrounding space was the only source of magnetic effects in the experiment.[8]

In Helmholtz's time, theoretical and experimental investigations of a high level of refinement mutually supported one another and were made possible by improvements in instrumentation. In fact, around the middle of century, electric instrumentation in steady or slowly varying currents and static potentials had reached a remarkable standard of precision in Germany and in the United Kingdom. In Germany Weber contributed by his invention in 1841 of the electrodynamometer, which he used in 1846 to test his induction law, and in 1846 of the mirror galvanometer. Kirchhoff advanced instrumentation techniques in his studies in 1845 of the distribution of currents in wires, as did Weber through his exact determination of electric units in 1852. Weber's determination had international repercussions: in 1861 the Royal Society appointed a commission headed by William Thomson to fix the standard unit of resistance, the so-called "Weber". Weber and Kohlrausch's application of the electrodynamometer in 1855 to determine the electrodynamic unit of current was inspired by the theoretical approach to electrodynamics suggested by Weber's elementary law of force.

Helmholtz discussed theoretically in 1847 the oscillatory nature of a condenser discharge, and in 1851 and 1869 he detected oscillations in open circuits.[9]

William Thomson treated condenser oscillations in 1853, presenting the well-known relation between period, capacity, and inductance; and W. Feddersen measured periods of oscillations by his mirror technique between 1858 and 1862. Theoretical problems such as those related to the inertia of charges encouraged physicists to take an interest in oscillations of higher frequencies and, in turn, their interest carried over to propagation phenomena in wires.

Following this research, an important experimental subject for Hertz, the propagation of currents in wires, became amenable to scientific study.[10]

This short review clearly shows that, in the 1870's, Electrodynamics was a highly mathematised discipline. German electrodynamics was still developing along the lines laid down by the mathematical physicists of the middle of the century: Franz Neumann. Wilhelm Weber, and Gustav Kirchhoff. In the same years, following George Green's, Thomson's and Maxwell's original approaches, British electromagnetism was developing with the aid of the mathematical tools of elasticity, hydrodynamics, and thermodynamics.

5.2. Prior to Hertz : Experiments on Distant-Induction, not a Test of Maxwell"s Electromagnetic Theory

I will now examine Maxwell's theory from the point of view of its experimental confirmation, looking specifically to see if it suggested any experimental approach in the range of frequencies in the "electromagnetic band". One of the main pieces experimental evidence Maxwell presented in support of his theory was his law for the speed of light $c = 1/\sqrt{\varepsilon\mu}$, where ε and μ are respectively the "specific capacity for electrostatic induction" and the "magnetic permeability". As another piece of evidence, he considered the Faraday effect concerning the rotation of the plane of polarisation of light in a magnetic field.[11] Maxwell's selection of evidence confirms the view that his strongest interest in testing the theory lay in the domain of the electromagnetic theory of light.

As regards a possible test of his theory in the "electromagnetic band", the following remarks should be made. The whole first volume of Maxwell's *Treatise* deals with electrostatics and steady currents, and only a limited number of pages of the second volume deal with the theory of the electromagnetic field. In the second volume no trace can be found of a theory of electric oscillations, and in the first volume condenser discharges are treated as an aperiodic phenomenon.[12] More important, perhaps is the absence in Maxwell's writings of any theory connecting a propagating field and an oscillating current as its source; his well-known solution for plane waves[13] corresponds to the case of a source at infinity and a vanishing d'Alembertian of the field.

These remarks illustrate that experiments for testing the theory in the "electromagnetic band" were not immediately foreseeable. Indications of a possible detection of the displacement current were, however, given by Maxwell. In 1868 Maxwell pointed to the detection of the displacement current "within the dielectric itself by a galvanometer properly constructed" as a possible way of proving the theory. The displacement current, which Maxwell considered "a natural consequence" of his theory, was "not yet verified by direct experiment ... [which] would certainly be a very delicate and difficult one".[14]

Given Maxwell's manner of presenting his theory, it is not surprising that an experimental test of the "electromagnetic band" would have been possible only in a wholly different theoretical and experimental context. In effect, all experiments on free propagation prior to the 1880's were extraneous to Maxwell's theory, indeed to any established theory.[15] In 1875 Thomas Edison noticed sparks from metallic objects in the vicinity of a magnetic vibrator relay. In 1871 and after, Elihu Thomson noticed the same in the vicinity of a Ruhmkorff coil; with proper adjustments he was able to detect sparks as far away as the sixth floor of an observatory. Arnis E. Dolbear received signals in a telephone receiver in 1882, as did David E. Hughes in 1879. These effects were interpreted variously at the time as induction effects, or the manifestation of a new principle or a new "etheric" force "that was as distinct from electricity as was light or heat". Dolbear, Edison, and Hughes had very limited scientific backgrounds, and none was likely to be familiar with partial differential equations. Later Thomson and Edison regretted that they had missed a practical invention in wireless communication, not an advancement of science. Hughes, who thought that he had discovered a new principle, was discouraged from further inquiries by the opinion of such a distinguished physicist as George Gabriel Stokes that he was observing ordinary induction effects.[16]

All of these experiments were qualitative, especially since what impressed the experimenters was solely the distance over which the "induction" was detected. The distance effect was meaningless in the absence of a theory of the source-field relation.

5.3. Hertz's Conciliation of Two Distinct Research Traditions: Electric Waves in Wires and in Space

Faraday's induction law posed a theoretical problem within the two research traditions that are most relevant to Hertz's electrodynamic contributions: theories and experiments on conduction in wires on the one hand, and theories of interaction of forces and propagation of waves on the other.[17]

Prior to Hertz and Heaviside, neither Thomson[18] nor the German physicists related the propagation of conduction current in wires to a conception of the propagation of transversal waves in the space surrounding the wire or to any theory of the electromagnetic field. The propagation in wires was understood until 1885 as a propagation of waves of current within the wire.[19]

This understanding is illustrated by Wilhelm von Bezold's research on oscillations in 1870. Bezold produced oscillations by using a Ruhmkorff induction coil, whose spark-gap was connected to an aerial wire, which was the seat of propagation phenomena such as standing waves. To detect a propagation effect, Bezold used the technique of the so-called Lichtenberger dust-figures, based on the appearance of a compound powder in the vicinity of either positive or negative charges. The presence of "positive" figures where he expected "negative" ones suggested to him characteristic oscillations of charge. He also initiated the method of balancing the difference of potential on a sparking micrometer by different lengths of the connecting wires, a method which Hertz would fully investigate and exploit. Bezold argued that "phenomena occurred in electrical disturbances similar to those which are observed in the motion of fluids under the name of aspiration phenomena".[20] He spoke of "waves of electric discharge" or "electricity rushing along a shorter path", clearly showing that he saw oscillatory phenomena as a manifestation of moving charges or displacement of currents, unlike Hertz's "waves of potential". His experiments on the propagation of currents show that Bezold belongs to the group of physicists working within the tradition of research on propagation in wires.

As a consequence of their restricted view of current propagation, physicists before Hertz and Heaviside did not formulate in clear mathematical form their ideas concerning propagation and their conceptions of propagation of electric actions in free space. There is evidence that physicists who elaborated theories of propagation in wires, like Weber and Kirchhoff in Germany and Thomson in England, were either not in the least concerned with free propagation or did not stress any connection between it and propagation in wires. Thomson not only concluded that there was no definite velocity of propagation in cables or lines, but that there was also none in air and ether. Maxwell, too, made a distinction in his *Treatise on Electricity and Magnetism* between cases of diffusion, as in heat flow, and cases of propagation with definite velocity. The former applied to cables, the latter to free propagation of electromagnetic waves.[21]

A major field of theoretical electrodynamic research in the 1870's was the construction of theories of retarded action. Two theories of retarded action were developed in Germany by Riemann and Carl Neumann. Riemann's remained almost ignored; he had immediately withdrawn his paper, having mistakenly attempted to prove that the retarded scalar potential of the charge elements in two conductors is equivalent to the electrodynamic potential derived from Weber's law.[22]

Carl Neumann, a follower of Weber, derived in 1868 a theory of delayed action at a distance independently of any conception of an ethereal transmission of force. He expressed the potential energy between two electrical particles in motion in terms of a time lag, defined by the distance between the particles at the time when the action from the first particle reaches the second divided by the conversion factor c_W. Neumann believed that the interaction between events separated in time was primary, and not further explicable. He considered his concept of "retarded action" as a "transcendental concept", and electrodynamic action as basically different from the transmission of light or heat.[23] The Danish physicist Ludwig Lorenz also published an important theory of retarded action, based on Kirchhoff's equations for continuous three-dimensional currents. Hertz's theory of 1884 led to results having a formal resemblance to those of Lorenz, as Hertz explicitly noted.

Maxwell's theory of the electromagnetic field, which he developed between 1856 and 1873, grafted its conceptions of free propagation onto Faraday's dielectric conceptions, which wavered between the different views of dielectric action as polarisation of space-filling matter and as independently existing lines of force.[24] A characteristic feature of Maxwell's theory was the attention it paid to the "medium", whether by attempting to imagine a "mechanism" of transmission of electric action or by simply stressing the "dielectric" property of ether and its capability of sustaining "electric displacement". Electromagnetic energy had its seat in the space surrounding the conductors, a feature that Thomson had started to develop even before Maxwell's memoirs of 1862 and 1865. Maxwell understood and rejected continental ideas of currents and charges but his own ideas concerning currents and charges were anything but definitive and clearly expressed.[25]

Maxwell was thus unlikely to contribute to the theory of propagation in wires. Nor were the theoreticians of "retarded action" like Riemann and Carl Neumann in a better position. A possible exception is Lorenz, who derived an equation for current propagation from Kirchhoff's theory, but who soon identified the current density vector with the well-known free transverse light vector of the elastic theory of light; he also introduced a retarded vector potential into his theory. Riemann, Carl Neumann, and Lorenz were not interested in theories of propagation in wires and made no significant contribution to them.

The gap that existed between the conceptions and theories of the two modes of propagation was due, I believe, to their different historical development. Theories of circuits were mainly developed by experimentalists with mathematical and instrumental tools that were proper to stationary phenomena. Theories of retarded action, on the other hand, were mainly correlated with optics and elasticity, and they therefore required a more sophisticated mathematical approach. Moreover, the instrumentation suitable for detecting free propagation was in a primitive state.

It was Heinrich Hertz who developed suitable instrumentation to show that waves of current and waves in space interfere with one another, and who elaborated theoretically the mutual relations of the two modes. Studying in Berlin, Hertz had the opportunity to attend Helmholtz's and Kirchhoff's lectures and seminars and to work in Helmholtz's laboratory. He was well acquainted with Helmholtz's theory and William Thomson's theoretical and experimental advances in understanding propagation in cables. In an 1884 paper he also mentions the theories of Riemann and L.V. Lorenz, and he had studied thoroughly the relation of Helmholtz's theory to

Maxwell's theory of free propagation in ether.

Hertz was in an excellent position to unify experimentally and theoretically the two research traditions. He reached a synthesis first by an original theoretical approach in 1884 and then by his audacious experiments and theory four years later.

So far my main task in this section has been to outline the two research traditions leading to their unification by Hertz. Other points I wish to touch on in the following pages deal with the difficulties that had to be met and the price paid for the unification. The difficulties arose because the unification was achieved in the range of frequencies of what is now called the "electromagnetic band." As I have shown, in this range Maxwell's theory led nowhere and Maxwell's conceptions were peculiarly vulnerable. The price of unification was the need to abandon the concept of force in its Newtonian-Helmholtian interpretation. It was Antoon Lorentz's task to again reinstate force in Electrodynamics.

5.4. Helmholtz's Electrodynamics

In 1870 Helmholtz began publishing a series of articles that together constituted a comprehensive study of electrodynamics. He approached the problem of action at a distance and contiguous action in an original way by combining Poisson's theory of dielectrics with Franz Neumann's potential theory to yield a Maxwell-type theory of propagation through a "medium". Helmholtz's first article in the series, "On the Equations of Motion of Electricity for Conducting Bodies at Rest",[26] to which Hertz often referred in the course of his work, was a vast "tour d'horizon" of the competing theories of electrodynamics. It is significant that only sixteen out of the eighty-four pages of this large article were devoted to a theory of dielectric action; the remaining part was concerned with a form of potential theory and its consequences for induction and the motion of electricity in extended conductors.

Helmholtz started with an electrodynamic potential of two elements of circuits $d\sigma_1$, $d\sigma_2$ at a distance r from one another and carrying currents of intensity u_1, u_2. In modern vector notation,[27] the potential is:

$$\frac{A}{2} \frac{u_1 u_2}{r} \left[(1+K) \, d\sigma_1 \, d\sigma_2 + (1-K) \frac{r \, d\sigma_1 \, r \, d\sigma_2}{r^2} \right] \quad ,$$

where A is Weber's ratio or conversion coefficient from electromagnetic to electrostatic units. The different values of K define the potential functions of Weber (K=-1), of Franz Neumann (K=1), and of Maxwell (K=0). When Helmholtz's potential is integrated around a closed circuit, any dependence on K is lost, so that only experiments with open circuits seemed suitable for discriminating among the competing theories. Helmholtz extended his potential for linear currents at positions p and p' to volume currents: the potential per unit volume of a current of density u at the position p and time t is:[28]

$$-A^2 U(p,t) \cdot u(p,t),$$

where:

$$U(\mathbf{p},t) = \int_?^? \left[\frac{1+K}{2} \frac{\mathbf{u}}{r} + \frac{1-K}{2} \frac{\mathbf{r}}{r^3} (\mathbf{r} \cdot \mathbf{u}) \right] d\tau$$

$$\text{and,} \qquad r = p - p' \qquad .$$

From this potential, Helmholtz derived a theory for the motion of electricity in extended conductors, yielding the possibility of the propagation of longitudinal "electric waves" in a conductor. (This derivation was independent of the introduction of dielectric action, so that on this point Helmholtz was on the same ground as Kirchhoff and Weber, who did not connect propagation in wires with free propagation.) The experimental detection of longitudinal waves and the determination of K met with theoretical limitations; Helmholtz deemed experiments possible only for conductors whose transverse dimensions were very large in comparison with the wavelength, a case which was then hardly testable.

In Helmholtz's treatment of theories of dielectric action, he regarded the polarisation of dielectrics in Poisson's sense as resulting not only from static forces but also from electromagnetic ones; i.e., he recognised an additional polarisation produced by the time variation of an electric or magnetic field. His mathematical treatment of polarisation begins with the definition of polarisation in the static regime,[29] $P = \varepsilon (x - grad\ \phi)$, where ε is the dielectric constant, ϕ the electric potential function of the distributed electricity, and x the electric force. The polarisation in the dynamic regime is:

$$\frac{P}{\varepsilon} = - grad\ \phi - A^2 \frac{dU}{dt} + \frac{d}{dt} rot\ L + X \qquad ,$$

where **L** is the magnetic vector potential and **X** the external force. Here:

$$L = \int \frac{\lambda}{r} d\tau \qquad ,$$

where λ is the magnetic moment and is defined by $L = \theta(Q - \text{grad} \chi)$ with θ the magnetic susceptibility, χ the magnetic scalar potential, and **Q** the external magnetic force. From the equation for the electric polarisation and the corresponding equation for the magnetic moment, Helmholtz deduced an equation for the propagation of **P** when **X**=0:

$$\nabla P = 4\pi\varepsilon(1 + 4\pi\theta) A^2 \frac{d^2 P}{dt^2} + \left[1 - \frac{(1+4\pi\theta)(1+4\pi\varepsilon)}{K}\right] \text{grad div } P,$$

One finds solutions for the wave equation by considering **P** as composed of two vectors P_1 and P_2 such that rot P_1=0 and div P_2 = 0. Since the longitudinal solution belongs to that component of **P**, say P_1, for which rot P_1=0 and div P_1=1/4π grad div j , this component is an irrotational polarisation dependent, according to Helmholtz, on the static force. I will call P_1 a static-type polarisation indicating that this qualification refers to its spatial distribution and distance dependence, quite apart from its constancy in time. The velocities of propagation for P_1 and P_2 are, respectively,[30]

$$\frac{1}{A} \frac{1}{\sqrt{4\pi\varepsilon(1+4\pi\theta)}} \quad \text{for transversal waves,}$$

$$\frac{1}{A}\sqrt{\frac{1+4\pi\varepsilon}{4\pi\varepsilon K}} \quad \text{for longitudinal waves.}$$

It is especially important to mention here one of Helmholtz's subsequent areas of research: in 1874 he extended Franz Neumann's potential formula of 1848 to three-dimensional conductors and to the case of open currents, showing that from this extended potential ponderomotive as well as electromotive induction forces could be derived. The ponderomotive forces were different from those predicted by Ampere's formula, for they included also ponderomotive and inductive forces due to the charges situated at the ends of open circuits[31]. Hertz later considered these forces emanating from the open ends of a circuit to be an explanation for the failure of his early experiments on the electromagnetic effects of material dielectrics.

5.5. Helmholtz's Theory of Dielectric Polarisation

The basic tenet of Helmholtz's polarisation theory of electrodynamics is the conception of a "bare" charge in a vacuum or "empty space", as distinguished from a dielectric space or ether. He often qualified the word "ether" with the word "light". He considered the hypothesis of the polarisability of the light-ether as a tentative extrapolation from the dielectric polarisability of some material insulators and said that once the light-ether was considered magnetisable, the "moment is no longer far off when one can consider it also as a dielectric in Faraday's sense".[32] He related the "bare" charge of density E in a vacuum to the two potential functions which gave rise to a force in a vacuum, and f, which depended on the polarisability of the ether:[33]

$$-\frac{1}{4\pi}\nabla(\psi+\varphi) = \frac{E}{1+4\pi\varepsilon_0},$$

where ∇ is the Laplacian operator, ε_0 the dielectric constant for ether-filled space or, equivalently, air. It is a property of the whole potential function that in a space where E exists, $\psi+\varphi$ "behaves as if only $\frac{E}{1+4\pi\varepsilon_0}$ would be present in a non-dielectric space". As a consequence of this property any actual measure of charges by Coulomb forces would give $\frac{E}{1+4\pi\varepsilon_0}$. The bare charge E can only be measured together with the factor $\frac{1}{1+4\pi\varepsilon_0}$, which is indeterminate owing to the unknown value (one cannot remove the ether and measure the force between bare charges).

For Helmholtz, the unit of the electrostatic charge, as experimentally determined by Coulomb's law, is affected by the same indeterminacy of unknown multiplicative factors. Consequently, Weber's ratio A, defined as the ratio between bare electrostatic and electromagnetic units of charge, differs from c, the velocity of light in air, by $1/A = c\,(1+4\pi\varepsilon_0)(1+4\pi\theta_0)$, where θ_0 is the magnetic susceptibility of the ether.[34] Helmholtz developed an argument for θ_0 parallel to that for ε_0. The discrepancy between the "true" velocity $1/A$ and the observed velocity c in air or ether entails a correction in the theoretical velocities of longitudinal and transversal waves of polarisation **P**. The corrected values are:

$$c\left(1+4\pi\varepsilon_0\right)\sqrt{\frac{1+4\pi\theta_0}{4\pi\varepsilon_0 K}} \quad \text{for longitudinal waves,}$$

$$c\sqrt{\frac{\left(1+4\pi\varepsilon_0\right)}{4\pi\varepsilon_0}} \quad \text{for transversal waves.}$$

According to Helmholtz, the longitudinal waves of **P₁** were dependent on the static-type force. It is my opinion that Helmholtz's admission of a static-type force in the variable regime is to be attributed to his paradigmatic choice to consider "Kraft" in its substantive relation to matter. In his 1847 memoir on the conservation of "Kraft", for example, he affirmed: "It is evident that in the application of the ideas of matter and 'Kraft' to nature, the two...should never be separated".[35] I think that for him the counterpart in electrical science was the substantive relation between electric charge and electrostatic force.

The relation was tied to the nature of electric charge and therefore could not be dissolved by accidents such as the motion of the charge.

Although Helmholtz presented a wave equation for the polarisation, his physical ideas were by no means identical with Maxwell's. Helmholtz was aware of the differences, pointing out that his and Maxwell's theories

> are opposed to each other in a certain sense, since according to the theory of magnetic induction originating with Poisson, which can be carried through in a fully corresponding way for the theory of dielectric polarisation of insulators, the action at a distance is diminished by the polarisation, whereas according to Maxwell's theory the action at a distance is exactly replaced by the polarisation.[36]

Helmholtz's model of polarisability modified Maxwell's theory of the electromagnetic field in major ways: in Helmholtz's modification not only were two distinct forces present, but the velocity of neither coincided with the velocity of light. Since this was a consequence of his model of polarisability, Helmholtz had to force his theory to yield the velocity of light. For a very large value of the dielectric constant the velocity of the transverse wave converges on c. Moreover, the velocity of the longitudinal wave becomes infinite for K=0, in which case the longitudinal force acts at a distance. This condition also affects the intensity of the longitudinal force, as Helmholtz showed elsewhere:[37] in the case of very large values for the electric and magnetic susceptibilities, ε_0 and θ_0, the longitudinal distance force vanishes together with the free electricity E. The very large values of the susceptibilities do not effect the total charge, which remains finite.

The idea that distance forces emanating from bare charges exist in a dynamic system in a vacuum was deeply enmeshed in Helmholtz's theory of polarisability. Two consequences were that longitudinal, static-type forces are present in dynamical phenomena over the entire band of frequencies, and that a correction must be applied to the velocity of propagation of transverse and longitudinal forces as given by the d'Alembert equation of polarisation. Thus, an etherless vacuum and action at a distance had a conceptual primacy in Helmholtz's theory.

Helmholtz's theory was widely influential. Hertz took it as the starting point for his research in 1887, as did Lorentz, who accepted action at a distance in the Helmholtian formulation as the basis for his investigation of Maxwell's electromagnetic theory of light. Henri Poincaré devoted many pages of his *Electricité et Optique* of 1901 to an assessment of Helmholtz's theory and its relation to Maxwell's.[38] Pierre Duhem, after strongly criticizing Maxwell's theory and Boltzmann's formulation of it, recommended Helmholtz's theory as "a natural continuation of the theories of Poisson, Ampere, Weber, and Neumann," establishing a "continuity of tradition, without missing any of the recent conquests of electrical science".[39]

Helmholtz's theory has received attention recently from historians of science. Leon Rosenfeld, in his provocative essay, "The Velocity of Light and the Evolution of Electrodynamics," says that not only was Helmholtz's approach to Maxwell's theory "entirely alien to its spirit, but it tended to obscure its characteristic features and to make the theory appear as a somewhat singular limiting case of the scheme".[40] I agree with Rosenfeld's judgement; for I believe that there is no contradiction in the assertions that Helmholtz was establishing a "continuity of tradition" in the context of continental physics and that he did violence to Maxwell's ideas as they were understood by Maxwell and by some of his successors. I explain this seeming paradox by the differences between continental and British science; the former stressed the internal consistency and phenomenological concreteness of theories, whereas the latter stressed models and modes of representation. Helmholtz's interpretation of Maxwell conformed to the continental ideas.

Helmholtz's 1870 article, in which he placed the different electrodynamical theories on a basis that allowed for a decision between them, was consistent with Helmholtz's general ideas on scientific inquiry. He believed that

the impact of a new abstraction [such as the concept of electric displacement] can only be understood clearly when its application to the chief group of individual cases which it comprises has been thought out and found valid. It is very hard to define new abstractions in universal propositions, so as to avoid misunderstandings of all kinds.[41]

Helmholtz's concrete interpretation of Maxwell's displacement current as the dielectric polarisation of insulators is, perhaps, the best exemplification of this frame of mind.

5.6. Helmholtz's Experiments on Maxwell's Theory

Other experiments on the propagation of "induction" were performed in 1869 and 1871 in Helmholtz's Berlin laboratory, but this time in connection with a well-defined quantitative problem: how can one measure the velocity of propagation of "induction" or, at least, a lower limit to it? Using a pendulum as a current-switch, Helmholtz measured, with the precision of 1/231,170 seconds, the time lag between the opening of a primary and of a secondary coil. When the coils were 136 centimetres apart, a change in the delay of 1/231,170 seconds did not produce any change in the induced charge. Helmholtz interpreted this result as evidence that "if the induced action propagates with a finite velocity, this should be larger than 314,400 meters per second".[42] The experiments were situated in a theoretical context of not very clearly defined contours, as they were inspired indiscriminately by theories of retarded action including Neumann's, or by such ideas about propagation as those championed by Faraday and Maxwell. In effect, Helmholtz saw in this experimental approach a possible refutation of Weber's action-at-a-distance theory. Helmholtz appreciated the intrinsic limitations of a mechanical switch and the impossibility of sharply defining the initial and final values of current pulses for such a short interval of time.[43] He soon abandoned this method in his search for an experimental decision between the two main conceptions of electrodynamics; he turned instead to the indirect effects of finite propagation such as the polarisation of dielectrics.

    I need now to assess the role of Helmholtz's theory in inspiring experimental activity in his Berlin laboratory in the 1870's and, in particular, inspiring the experimental approach to dielectrics from which Hertz began his experimental study of electromagnetic radiation. I will begin with a brief description of the experimental problem and will follow that with a more detailed appraisal of the Helmholtz-Hertz relationship. Mechanical motion, in the form in which it had been exploited in induction-type experiments, seemed at first to Helmholtz suitable for producing those variations of electric and magnetic forces to which the polarisation of dielectrics was related in his formula.

Accordingly, Helmholtz limited the major part of his experiments in the 1870's to testing polarisation effects in the vicinity of charged conductors in motion. An experiment using a cylindrical condenser that turned around its axis in a uniform magnetic field was performed by N. Schiller in Helmholtz's laboratory and presented by Helmholtz in a memoir to the Berlin Academy in 1875.[44] Schiller concluded that Franz Neumann's generalised law of induction derived from his potential theory was false, whereas the results of the experiment agreed with Neumann's potential theory when augmented by a dielectric ether, and they agreed with his electromagnetic induction law derived from Ampere's action-at-a-distance forces and Faraday's dielectric polarisation (Helmholtz regarded Neumann's potential theory augmented by a dielectric ether as equivalent to Maxwell's theory).

An experiment that Helmholtz performed with Henry A. Rowland and that he reported to the Academy in 1876 confirmed that charges borne by moving material bodies exert, under certain conditions, magnetic actions[45]. But these magnetic actions could be interpreted as due either merely to the displacement of charges through the motion of their ponderable carriers, as in Weber's theory, or to the variation of the dielectric polarisation of air, or ether, in a fixed volume of space resulting from the motion of the electric force.

To discriminate between these two possibilities, experiments were required in which polarisation effects could be produced without the motion, at least the macroscopic motion, of charges. Helmholtz proposed such experiments as the theme for a research competition at the Academy in 1879, one that Hertz was to enter (see next chapter). Helmholtz's proposed experiments had the advantage, essential to experiments on electromagnetic waves, of requiring no mechanical motion to produce the expected effects; Hertz's new technique of detecting electromagnetic effects by purely electromagnetic means such as scintillation in a spark gap followed Helmholtz's line of thought.

CHAPTER 6

HERTZ'S EXPERIMENTS ON ELECTROMAGNETIC WAVES

6.1. Hertz's Initial Helmholtian Approach to the Experiments on the "Exceeding Mobility of Electricity"

Heinrich Hertz's decision to go to Helmholtz and to work in his laboratory in Berlin had great significance for his career in research. Helmholtz's general outlook on science and physics had an initial bearing on Hertz's own.

In the autumn of 1878 Hertz went from the University of Munich to the University of Berlin, where he spent his second year as a physics student.[1] In Berlin he followed Kirchhoff's and Helmholtz's lectures and attended Helmholtz's physics laboratory. He was soon drawn into Helmholtz's scientific circle; when shortly he undertook original research on a subject proposed by Helmholtz, the latter provided him with research facilities in his institute and paid daily attention to Hertz's progress. In 1880, Hertz published his results under the title "Research to Establish an Upper Limit for the Kinetic Energy of Electric Current",[2] for which he was awarded a prize by the Berlin Philosophy Faculty. Since this research was representative of the approach to electrodynamics common in Helmholtz's laboratory an approach that Hertz was to abandon eventually I shall comment briefly on it.

Hertz set out to prove experimentally that only a limited fraction of the extra current in a self-inductive circuit was dependent on the inertia of the current. He also sought an upper limit for the density of mass per unit charge. The equation for an oscillating current in a circuit with inductance is:

$$Aidt = i^2wdt + d(Pi^2) \quad ,$$

where A is the electromotive force, i the intensity of the current, w the resistance, and P the self-induction coefficient ("Potential des Leiters auf sich selbst"). Introducing the hypothesis of an inertial mass m of the current, this equation reads:

$$Aidt = i^2\,wdt + d(Pi^2 + mi^2) \quad .$$

In this case P should represent that part of the self-induction that is independent of inertial effects. All inertial effects are included in m = r qll, where l is the unit of positive electricity contained in unit volume of wire, l the total length of wire, q the cross section of the wire, and r the density of mass per unit charge.

Hertz sought to isolate the inertial element m by comparing the inductive effects in a series of circuits with varying self-induction.[3] He measured the extra current with a galvanometer inserted into an arm of a Wheatstone bridge. The mechanical arrangement made it possible to establish contact with different circuits in quick succession. At first Hertz used, on Helmholtz's suggestion, double-wound or Knochenhauer spirals to reduce the self-induction. However, he soon found out that straight wires in circuits of a rectangular shape gave better results, since the calculation of the self-induction coefficient was then geometrically simpler. (Both rectangular circuits and self-induction calculations would enter his later research on electric waves.) Assuming for the velocity of the current the values 1 mm/sec and 10 mm/sec, he obtained from the experiment the corresponding inequalities:[4]

r < 0.008 milligrams ;   r > 0.00008 milligrams.

His conclusion was that the upper limit for the density of mass seemed to disprove Weber's electrodynamic law; for the small, if not vanishing, inertial effects implied an instability in the charge distribution of a metallic sphere.[5] This argument had been developed by Helmholtz in his 1870 essay, showing clearly that Hertz's first research belonged to the terms of the Weber-Helmholtz debate. Helmholtz confirmed this point in his 1894 reconstruction of Hertz's route to the discovery of electric waves. There Helmholtz stressed the importance for Hertz's future work of his early recognition of the quasi absence of inertial effects in the motion of electricity:

These experiments clearly impressed upon his [Hertz's] mind the exceeding mobility of electricity, and pointed out to him the way towards his most important discoveries[6].

Helmholtz's polarisation theory stimulated Hertz's interest in a conception of contiguous action in electromagnetic phenomena, although Hertz's early approach took its starting point in the Riemann and Carl Neumann tradition of retarded action. It is clear from Hertz's writings that he had studied Helmholtz's theory with its peculiar combination of distance and contiguous action[7].In 1879, on Helmholtz's suggestion, the Academy offered a prize for the solution to the following problem(hereafter, Academy prize):

To establish experimentally any relation between electromagnetic forces and dielectric polarization of insulators- that is to say, either in electromagnetic [electrodynamic] force exerted by polarisations in non-conductors, or the polarisation of a non-conductor as an effect of electromagnetic [electrodynamic] induction.[8]

At that time, Hertz did not think that the oscillations of Leyden jars or open induction coils would lead to observable effects.[9] He did not say why he thought the effects were unobservable, but his reason can be inferred from a note in his diary,[10] and, especially, from the content of his 1887 paper, "On Very Rapid Electric Oscillations," his first response to the Academy's problem. He felt that he needed more rapid oscillations than ordinarily obtainable, since any dynamical polarisation effect was dependent on time derivatives.

In 1883 Hertz moved from the University of Berlin to the University of Kiel and from his position of assistant to Helmholtz to that of Privatdocent for mathematical physics. His outstanding theoretical work at Kiel was his 1884 paper, "On the Relations between Maxwell's Fundamental Electromagnetic [Electrodynamic] Equations and the Fundamental Equations of the Opposing Electromagnetics [Electrodynamics]".[11]

Since the evaluation of Hertz's 1884 contribution has been highly controversial in recent historiography, I prefer to devote the rest of chapter six to an examination of Hertz's paper. I believe that the nature of Hertz's 1884 arguments and their connection with his 1888-90 contributions and especially with his magisterial 1888 experiment the subject of my next two chapters will be better understood when situated against the background of Hertz's subsequent research.

I have stated one aspect of the relationship hid between Hertz and Helmholtz. Their association, however, was many-sided. From the experimental point of view, it was decisive for Hertz's success that he approached Maxwell from Helmholtz's point of view. In fact, Helmholtz's program of detecting the polarisation of material dielectrics by induction experiments, which Hertz took up, implied circuital electricity; the dielectric body had to be inserted into the primary circuit to make its inductive effects observable. Helmholtz's circuital approach was an important factor in leading Hertz toward his understanding of the source-field relation. The stress on the electromagnetic forces of polarisation currents brought to Hertz's attention the more general connection between conduction currents and fields in the vicinity of a linear oscillator. Hertz generalised this connection to that of a relation between sources and waves in the radiation field of the same oscillator.

It will serve as an indication of the complexity of the historical situation to remark that Hertz's experiments, in their initial phase, turned out to be significant for Helmholtz's potential theory as well as for Maxwell's theory. The two theories were concerned with the distribution of forces in the neighbourhood of an unclosed current, and Hertz found a way to detect precisely that distribution. In fact, in a letter to Helmholtz dated January 21, 1888, Hertz spoke of the possibility of measuring K in Helmholtz's potential formula and his dielectric constant of space.[12] In the conclusions to his February 1888 paper communicated to the Academy on February 2 of the same year,[13] Hertz thought of using the new procedure to test these theories, especially the one which had received Helmholtz's imprint and sanction.[14] However, he showed no protracted interest in potential theories.

Hertz progressed towards his decisive experiments on electrical radiation by moving from one conception of electricity, in which charges and currents in circuits were the sources of force, to a different conception of electricity, in which the electric force was identified with the polarisation of the ether.

Hertz never returned in his mature work to the circuital

approach to electrical phenomena that typified his early research. The "exceeding mobility of electricity" was perhaps one of the first steps in his radical change in basic conceptions.

6.2 *From Conduction of Currents in Wires to Propagation of Changes of Potential*

In 1885 Hertz left the University of Kiel and accepted a position as professor of physics at the Karlsruhe Technische Hochschule, where he performed all of his experiments on electromagnetic waves. There in 1886 he experimented with the Riess or Knochenhauer spirals. These were double-wound wire spirals, a common electrical device for induction experiments that he had already used in his experiments on the inertia of currents in Berlin. He used the two spirals as induction coils of low self-inductance, inserted in a powered primary circuit and secondary passive circuit, respectively. He obtained strong sparks in the secondary coil when discharging the primary through a sparking gap. He attributed the strength of the sparks to the high frequency of oscillation in the spirals (paper no.1,"Introduction, A), Experimental")[15].

Hertz's transformation of the experimental set-up from the Riess spirals to sparking in the secondary circuit ("Nebenkreis") indicates his conceptual train of thought at the time. Arguing from the propagation of electrical effects in wires, he was trying to give a first evaluation of the frequency of the oscillations that produced the discharge in the primary circuit. This train of thought led him to modify the initial arrangement by connecting by wire the primary to a given point in the secondary circuit. In his experiment (Paper No. 2), he intended to verify his conjecture that, in the secondary gap, sparks "(show) more clearly ... that these disturbances run on so rapidly that even the time taken by electric waves in rushing through short metallic conductors [in the secondary] becomes of appreciable importance. For the experiment can only be interpreted in the sense that the change of potential proceeding from the induction coil reaches Knob 1 in an appreciably shorter time than Knob 2".[16]

It was an experiment on propagation in wires, but the conceptual context was now "the propagation of change of potential" in wires and in air, or waves of potential, not the propagation of currents or charges in wires as it had been in his former experiments on the inertia of currents. This circumstance might partially explain how little attention Hertz had paid until now to theories and experiments, such as Bezold's, on the propagation of "waves of current" in conducting wires.

There is evidence that Hertz was acquainted with theories of propagation of current in wires[17] though he did not explicitly mention them. He often assumed, for instance, that the velocity of propagation is independent of the resistance of the wire in wires of small resistance, and this is one of the main tenets of wire propagation theories.

6.3. A Clue for the Academy Prize: Propagation of very Rapid Electric Oscillations in Short Open Linear Wires

What Hertz thought was new in his experiment was the propagation of high frequency oscillating currents. He was surprised to learn after his article appeared that fifteen years before Bezold had produced effects equal to his own. Bezold had not attributed to them the same importance for electrodynamics as Hertz did, and his work had remained almost unnoticed in scientific circles. Hertz justified his oversight by pointing out that the external appearance of Bezold's paper had led him to think that it concerned only electric dust-figures.

Hertz now viewed propagation phenomena as a manifestation of the rapidity of oscillations, as is shown by the title of the paper, "On Very Rapid Electric Oscillations", in which he described the experiments. Accordingly, he viewed sparks as the manifestation of very strong potential gradients across the gap and, more generally, all over the secondary conductor. His view was also that a marked non-uniformity of potential distribution was a sign of propagation with very short wavelengths. It can be argued that he considered the above qualitative evaluation of the frequency of oscillation through the wavelength as a way of circumventing the impossibility of directly measuring the frequency by methods like Feddersen's for less rapid oscillations.

The phenomenon of oscillations in both primary and secondary circuits also interested Hertz because the oscillator was a short metallic conductor with open ends, an "open circuit", and short open rectilinear circuits had an important position in Helmholtz theory. A letter from Hertz to Helmholtz in this period stressed precisely this point.[18] Previously, according to Hertz, oscillations had been obtained in open coils and Leyden jars, but never in short metallic conductors.[19]

The high-frequency aspects were responsible for the modifications that Hertz introduced step by step in the primary oscillator and the secondary circuit, as can be clearly seen in the sequence of figures in his paper, evolving eventually into the now familiar open radiator and receiver. But this was not precisely Hertz's object at the time; rather, he was interested in the rapidity of oscillations in the primary circuit and in a procedure not only for detecting them but for evaluating their frequency through the secondary oscillations. At this stage Hertz studied the behaviour of the secondary circuit as manifested by sparks in the spark gap, soon convincing himself that the secondary circuit was the basis of a resonance phenomenon and learning how to tune it with the primary circuit to magnify the sparks.

At the same time he realised that sparking was a more complicated effect than he had expected and that causes other than potential gradients combined in producing it; some of the causes, such as electrostatic ones, could be eliminated by interposing a wet thread between the knobs of the micrometer (paper no. 2).[20]

Another cause of the complications was the "photoelectric" effect, then still unknown, which Hertz, with clear judgement, attributed to the presence of ultraviolet light from the primary discharge; to eliminate this cause, he undertook a special investigation of the phenomenon and published the result in a separate paper, "On an Effect of Ultra-Violet Light upon the Electric Discharge" (paper no. 4).

In the theoretical section of "On Very Rapid Oscillations", Hertz computed the half period T of the oscillation by Thomson's formula:

$$T = p/c \sqrt{P C} ,$$

where c is the velocity of light, C the capacity of either one of two large spheres of 15 cm radius, attached to the ends of a straight metallic wire of the primary oscillator, and P the self-induction coefficient of the wire of length 150 cm. (Since he computed P and C in electromagnetic and electrostatic units, respectively, he introduced the conversion factor $1/c$.) He found that $T = 1.77 \times 10^{-8}$ sec. In computing C, he considered the relative capacity of the spheres instead of the absolute capacity of one of them, an error[22] that Poincaré[23] remarked on in 1891 and that Hertz himself acknowledged.[24]

At this stage Hertz considered the new rapid oscillations as a promising effect for solving the problem of the Berlin Academy. He thought that it would be easy with the aid of these rapid oscillations to detect the electromagnetic effects of a polarisation current in a dielectric block of sulphur or paraffin. He was impressed by the presence of a position, or neutral point, of the gap in which sparks were either very feeble or absent. His first attempt at detecting the effects was to insert a dielectric block between the metallic plates of the condenser in the primary circuit and then to remove it quickly. He expected that through the displacement of the neutral point, the detector in the vicinity of the block would discriminate between the induction of the entire loop with the block present and the induction when the block was quickly removed.[25]

Given Helmholtz's understanding of polarisation, which was inspired by the static-type Poisson polarisation of material dielectrics, it was reasonable that Helmholtz and Hertz should regard dense, usually solid, dielectrics with a high dielectric constant as suitable for experiments. That the dielectric was a solid favoured its insertion into a circuit but prevented the detection of the effect in its interior, limiting this detection to locations situated in its neighbourhood. The experiment failed, since he noticed no change in the sparking in the secondary circuit when the block was removed.

He decided that he had attacked the problem too directly and that the various parts of the secondary circuit in which neutral points were present had first to be studied. Modifying the rectangular form of the secondary circuit to that of a circle, he studied the primary oscillations by moving the circle which, being symmetric, prevented the disturbing effects of the asymmetry of the previous square detector. He now introduced an optical device to observe in the dark very feeble sparks in the spark gap and he made the gap adjustable by means of a micrometer screw. The secondary circuit was mounted on a wooden base. By this arrangement, he discovered that other neutral points existed in different positions with respect to the primary circuit.

6.4. A Phenomenological Theory of the "Kreis" Detector

Hertz consequently developed a theory of the secondary circular circuit which he described in paper no 5.[26] It was a phenomenological theory of the electric forces that affect the displacement of the "neutral point", into which he introduced concepts from electrostatics and the theory of electromagnetic induction at a semi-quantitative level. His theory assumes that stationary oscillations are induced in the secondary circuit and that the position of the spark-gap always corresponds to a node of the current.[27] He discusses theoretically the position and magnitude of the external exciting electrical force when the spark-gap corresponds to the position of maximum sparking in the circular secondary detector. He assumes further that the oscillations in the secondary circuit, which most effectively produce sparks, are the fundamental ones and not overtones. Accordingly, a maximum amount of sparking occurs when the position of the spark-gap is such that the plane of the circle is parallel to the primary inductor and the gap lies along the vertical to the plane, passing through the primary and the diameter of the circle. In this case, the theory predicts that the electric force is tangential to the circle at the spark-gap and to the position of the circle diametrically opposite. The direction of the external electrical force can be theoretically determined by placing the plane of the circuit in a vertical position the primary is horizontal and by bringing the gap to the highest position and then turning the circle around a vertical axis until the sparks disappear.[28] The sparks that are sensitive to this rotation are produced by the electrostatic force whose direction can thus be determined. The electromagnetic force acts in every position along the tangent to the circle, and the sparks produced by it are not sensitive to a rotation of the gap.

Hertz also infers from his theory of the detector that at positions beyond three meters from the primary circuit only the electromagnetic type of force seems active in producing sparks, and he depicts approximate maps of the lines of force.[29] The theory of the "circle" Hertz developed in his paper no. 5 will soon allow him to discriminate between electrostatic and electromagnetic forces.

The motive of my discussion of Hertz's theory of the "circle" is to indicate the rather elaborate state of the theory. However, some parts of the theory, such as the assertion that the gap always corresponded to a node of the current, were rough approximations or controversial affirmations.[30]

From the point of view of an exact theory, Hertz's experiments were complicated by the following feature, which was first noticed by Oliver Lodge, who in 1888 was working on similar experiments. Hertz's radiators were strongly damped and consequently oscillated over a large band of frequencies, whereas the opposite was the case for the detectors, which were persistent vibrators oscillating with little damping over a very short band of frequencies.[31]

From the practical side, this complication was a happy circumstance because it made it easy for Hertz to tune his detectors to different wavelengths according to their size and shape. On the other hand, the damping was not so strong as to prevent any interference between primary and secondary reflected waves. He had recognised this advantage very early in his work.

I am now ready to discuss Hertz's reaction to the Berlin Academy Prize research. Through the experiment described in paper No 5 he had found that the presence of an insulator modified the positions of the neutral point; he believed he was ready to find a solution to the problem of the Berlin Academy, namely, showing that "an electromagnetic force [is] exerted by polarisations in non-conductors."

In the experiment described in paper no. 6[32], he attributed the above modification or, more precisely, the angular displacement of the neutral point, to the inductive effect of the polarisation current in the dielectric block when the block was placed as close as possible to both primary and secondary circuits. In short, he used the apparatus as a kind of "induction balance"[33] that displaced the neutral points by superposing induced currents from conduction or polarisation currents.

The experiment is, however, mainly qualitative: the induction effect produced by electrical disturbances in insulators is manifested through the angular displacement of the direction of the null point. Hertz shows that the induction effects produced by the presence of large blocks of insulators like paper, paraffin, and asphalt are of the same order of magnitude and sense as those produced by thin metallic plates. He does not give a quantitative evaluation of the electromagnetic effect of dielectrics and conductors, but refers only to a "very rough estimate" he made on the assumption that "the quantities of electricity displaced by dielectric polarisation must be as great at least as those which are set in motion by conduction in thin metallic rods".[34] In this estimate, he probably uses Helmholtz's 1870 expression for the density of polarisation current. He argues that he has shown an equal effect both with conductors, in which currents exert electromotive forces according to the accepted theory, and with non-conductors. According to him, the second part remains untested. The research, for which he was awarded the prize, was communicated to the Academy on 10 November 1887.

It is worthy to notice that, at this stage of his research, Hertz considered the inductive effect of polarisation currents in dielectrics as confirmation of the views of Faraday and Maxwell.[35] He believed, too, that his experiment solved the first part of the problem of the Berlin Academy; namely, the problem of showing that "an electromagnetic force [is] exerted by polarizations in non-conductors." He was awarded the Academy prize.

I think that Hertz's experiment, which he considered a solution for the problem of the Berlin Academy, presents enough evidence to let us conclude that, up to 19 November, 1897, he interpreted the dynamic polarisation of dielectrics as an affect that confirmed Maxwell's theory, but that, at the same time, he thought that this effect did not contradict Helmholtz's thesis and the Academy question. Thus, one can explain Hertz's interpretation as a consequence of the ambiguous role assigned to polarization in dielectrics by the two competing theories, a matter I intend to examine in the following pages.[36]

6.5. Hertz's New Conception in 1888: Waves of Electric Polarisation in Ether

Hertz's paper on electromagnetic effects in insulators was dated 10 November 1887; his next paper was presented to the Berlin Academy on 2 February 1888. In this period of about three months, Hertz's ideas underwent a remarkable change, which was to affect all of his following work. In brief, the propagation of electricity in air and ether, the central theme of his 1884 investigations that he had subsequently abandoned in favour of Helmholtz's approach, again became central to his thought, but, this time in connection with ether polarisation. At this point Hertz moved from a concern with polarisation-current effects in the neighbourhood of large blocks of material insulators (i.e., material dielectrics) to the understanding of free propagation as an effect of an ethereal polarisation. He never returned to the separate test of polarisation effects of these insulators, dismissing the phenomenon as marginal to Maxwell's theory.

To begin with, let us note that the relation between wire propagation and free propagation is central to Hertz's change of mind, when he finally decided in favour of the latter, i.e., propagation in air of electrical waves. Hertz's conceptual change is evident from the new type of experiment he described in his paper of February 1888, (paper no.7) reported to the Berlin Academy on 2 February 1888, with the title: "On the Finite Velocity of Propagation of Electromagnetic Actions".[37]

This experiment was the first that was planned to demonstrate propagation in air. The paper opens by stating that the problem of the existence of polarisations in air accompanying electric forces is "another question" than that of the polarisation "within insulators whose dielectric constants differ appreciably from unity." The shift is from the program of testing polarisation effects in material dielectrics i.e., of testing the first and second hypotheses to that of testing waves of polarisation in air, or in empty space (the two phenomena were equivalent in this context). The existence of waves in air would prove simultaneously the two first hypotheses of empty space. Hertz concludes by stating that if air polarisation exists, then "electromagnetic actions must be propagated with a finite velocity".[38]

In the initial experiment on propagation, Hertz presents an interference between waves in air and waves in wires, the wire system being coupled to the primary circuit by a "capacitive" coupling.[39] Hertz had reasons for this hybrid combination, which caused him many difficulties: in case the velocity of electrical waves in air was much greater than that of light, a result which would agree with Helmholtz's theory, it would be safer to measure it by comparison with the velocity in a wire. He rightly expected a lack of directionality in the dipole radiation that would hamper direct interference, and, most important, he was much better acquainted with propagation in wires than in air.

Whatever Hertz's reasons for the comparison, what is significant for us is that to conceive of the likelihood of an interference between air and wire systems he had to regard propagation in air and in wire as essentially akin to one another.[40]

In his first propagation experiment, Hertz succeeded in demonstrating propagation in air, though he had difficulty determining the velocity, and his doubts on the validity of Maxwell's theory were at their greatest. He remarked that "the resulting interferences did not succeed each other at equal distances, but the changes were more rapid in the neighbourhood of the oscillation than at greater distances". He explained this irregularity in the spatial distribution of the interferences along the wire by the supposition that the total force might be split into two parts, of which the one, the electromagnetic was propagated with the velocity of light, while the other, "the electrostatic force ... is propagated with an infinite velocity".[41] Besides, the ratio of the velocity in a wire to the velocity in air of the electromagnetic force was about two-thirds.[42] This result was inconsistent both with traditional theory and with Maxwell's theory; both predicted a velocity equal to the free velocity in air in a conducting wire of little resistance.

The key to resolving this dilemma was to measure independently the velocity in air as he did in the experiment performed shortly after February 1888 and reported in paper no.8 : "On Electromagnetic Waves in Air and their Reflection".[43] The background of this experiment is the following: while experimenting with his "kreis", Hertz had noticed that the sparks' behavior near the walls of the room seemed to indicate a reflection of waves. He set about exploiting the phenomenon for measuring wavelengths in air in a stationary wave system. This plan had the unique advantage of allowing the measurement of the wavelength in air to be made independently of that of the wavelength in the wire, and, assuming in air a velocity equal to that of light, of allowing the period of oscillation to be calculated from the wavelength in air. Hertz found the period to be 1.55×10^{-8} sec, and showed that this result compared favourably with that obtained by Thomson's formula, i.e., 1.4×10^{-8} sec. He obtained a result which was consistent with the former one.[45]

6.6 *A Reinterpretation of the Experimental Results in the Light of Maxwell's Theory*

In his paper no. 9, "The Forces of Electric Oscillations Treated According to Maxwell's Theory",[46] composed in the spring of 1888, Hertz firmly establishes the new theoretical context of Maxwell's theory through a comprehensive treatment of one important aspect, the source-field relation. This aspect is worth commenting on. In the paper, Hertz presents Maxwell's equations at the beginning in exactly the same form and with the same symbols as in his 1884 paper. He restricts their solution to the case of a source of waves in the shape of a rectilinear oscillator along the X axis of a cartesian system of coordinates. He derives the components X, Y, Z of the electric and L, M, N of the magnetic force in terms of a quantity, known today as the polarisation potential or Hertz's vector:[47]

$$X = -\frac{d^2\Pi}{dxdy} \qquad Y = -\frac{d^2\Pi}{dxdy} \qquad Z = \frac{d^2\Pi}{dx^2} + \frac{d^2\Pi}{dy^2}$$

$$L = A\frac{d^2\Pi}{dxdt} \qquad M = A\frac{d^2\Pi}{dxdt} \qquad N = 0 \quad ,$$

where A is, as usual, the reciprocal of the velocity of light. Π defines a function of the cylindrical co-ordinates r, z and time t which satisfies the d'Alembertian equation:

$$A^2 \frac{d^2\Pi}{dt^2} = \nabla \Pi \quad .$$

A suitable solution for Π, which is satisfied everywhere except at the origin, is:

$$\Pi = El\frac{\sin(mr - nt)}{r} \quad ,$$

where E is a quantity of electricity, l a length, r the length of the position vector, $m = \pi/l$ and $n = \pi/T$; here l and T are the wavelength and period, respectively, and are related by $l = t/A$. In the immediate neighbourhood of the oscillator, where r is vanishingly small compared with l, and mr is negligible compared with nt, one has : $\Pi = -(El \sin nt)/r$. The electric force components are given in this case as the second derivatives of a double point potential, which represents the static potential of an oscillator of length l.

By the use of the polarisation potential Hertz succeeded in integrating Maxwell's equations for the particular situation of the dipole source. His method represents a special case of an approach of more general significance to the solution of the equations with source terms: the method was later to be known as that of retarded potentials.

Hertz's introduction into the theory of this potential facilitates, among other things, the development of analytical expressions for the "dynamical lines of force" of the radiation field. Hertz plots neat diagrams of the lines for different values of time, which he reproduces as illustrations in the article. This is the first time that a radiation field is represented as a drawing.[48] Hertz's computation of the energy emitted in a half-period from his oscillator is reported[49] as 2, 400 ergs.[50]

Hertz's acceptance of Maxwell's theory meant a decisive rejection of his previous concepts, among them the distinction between static and dynamic forces on which almost all of his previous thought was based. His introduction of the potential allows him to write the component of the electric force in the equatorial plane, lying on the axis of the oscillation and at a long distance from it, in the form:

$$Z = Elm^3 \left\{ -\frac{\sin(mr-nt)}{mr} - \frac{\cos(mr-nt)}{m^2r^2} + \frac{\sin(mr-nt)}{m^3r^3} \right\}.$$

From Hertz's statement, "the resultant wave can in no way be split up into two simple waves travelling with different velocities", we can argue that his earlier way of explaining the interference in wires by considering the total force as composed of an electromagnetic force propagating with the velocity of light and an electrostatic force propagating "with greater and perhaps infinite velocity ... can only serve as an approximation to the truth".[51] Hertz now regards the splitting up of the force as meaningless in general and his new conception as a permanent achievement. Hertz's recognition of only one kind of force now raises the question whether or not Maxwell's theory leads to any irregularity in the special distribution of interferences. Hertz finds the agreement with the theory satisfactory for interferences he attributed to the electromagnetic force, but not for those attributed to the static force.[52] He treats wire propagation in an original way, establishing an equation in which the direction of incidence of lines of force into the wire is correlated to their velocity of propagation.[53] He confirms that in a good conductor the wave should propagate with the velocity of light, not with the smaller velocity of his previous results; thus the situation really allows no escape. He also finds obscurities in the velocity of propagation along the axis of crooked wires and spirals, which he expects to be the same as the velocity along a straight wire.

The fact that Hertz takes up wire propagation again after he has achieved air propagation confirms the importance he attributed to this aspect of electromagnetic theory. However, in view of his achievement of air propagation and his theoretical explanation of it, he does not regard the difficulties in wire propagation as an obstacle to the acceptance of Maxwell's theory. His conclusion at this time is:

In our endeavour to explain the observations by means of Maxwell's theory, we have not succeeded in removing all difficulties. Nevertheless, the theory has been found to account most satisfactorily for the majority of the phenomena; and it will be acknowledged that this is no mean performance. But if we try to adopt any of the older theories to the phenomena, we meet with inconsistencies from the very start, unless we reconcile these theories with Maxwell's by introducing the ether as dielectric in the manner indicated by v. Helmholtz.[54]

6.7. A Radical Change: from Charges and Currents in Wires to Waves and Fields in Space

To sum up, the research Hertz reports in "The Forces of Electric Oscillations Treated According to Maxwell's Theory" represents the achievement of propagation in air, the determination of its velocity, and the theoretical understanding of the source-field relation through Maxwell's equations. This achievement allows Hertz in turn to clarify the inconsistencies he left behind him, and he does it through a complete acceptance of the Maxwellian context.

Here is the way Hertz elucidates this point in the same paper. He had measured the velocity in air independently of the velocity in wires and had found it to conform to Maxwell's theory or, equivalently, to the limiting case of Helmholtz's theory. However, some aspects of the phenomenon were still disconcerting: the phase distribution in air seemed to point to an infinite velocity in the vicinity of the generator.[55] This inconsistency was similar to the one Hertz had already found in the case of wire propagation, and which he had explained by splitting the force. But now in the course of the experiment, Hertz advances tentatively a different explanation based on energy localisation in space:

In the sense of our theory we more correctly represent the phenomenon by saying that fundamentally the waves which are being developed do not owe their formation solely to processes at the origin, but arise out of the condition of the whole surrounding space, which latter, according to our theory, is the true seat of energy.[56]

Hertz seeks to explain the instantaneous propagation of force in the neighbourhood of the oscillator, i.e., the propagation with a velocity higher than that of light, on the grounds that electromagnetic energy is localised and spreads out from the space surrounding the conductors. Therefore, interferences are not to be measured as if the waves propagated from the oscillator. Hertz affirms the new context more and more forcefully, and partly supports his judgement by referring to Heaviside's and Poynting's elaborations of Maxwell's theory.

In the summer of 1888 he carried out the experiments on the guided propagation of waves[57] described in his paper no. 10. In the following months, after completing his final experiment on free radiation, he resumed the experiments on wires,[58] an indication of the importance he attributed to them.

The view he was testing and that he believed was confirmed by his research was the one he ascribed to Heaviside and Poynting; namely, that "the electric force which determines the current is not propagated in the wire itself, but under all circumstances penetrates from without into the wire". Hertz considered it as "the correct interpretation of Maxwell's equations as applied to this case".[59] In his initial conception of guided waves, charges in the wire could still have the function of sources of external electric force, but in this last conception the force penetrates from without and determines the current.[60]

Hertz's statement presents impressive evidence for the conceptual change in the context within which he was working; from a context of charges and currents he passed to the comprehensive context of the field. The concept of the field now challenged the earlier concepts that had been firmly established on the continent for nearly a century. Hertz does not give a definite opinion on the locus of the disturbances in the case of a regime of steady currents. However, he also did not make a distinction between the variable and steady regimes in this connection; to have done so would have seemed inconsistent on his part after having had asserted the principle of the unity of electrical force.

At some stage in the researches he was working on in March 1888, Hertz discovered through his detector that his apparatus was capable of producing shorter wavelengths. His clear theory of the source-field relation now allows him confidently to shape his experiment to demonstrate the ondulatory nature of the electrical force. He concentrates electric waves with mirrors, refracts them with lenses and prisms, polarises them with metal gratings, and reports on all this in paper no.11, "On Electric Radiation" ["Uber Strahlen elektrischer Kraft"],[61] which he presents to the Berlin Academy on 13 December 1888. In place of the large Ruhmkorff coil of the previous experiments he uses a smaller induction coil. At times he substitutes for the circular detector the now well-known dipole antenna; the antenna is 160 centimetres long and is not at this time synchronised with the radiator. He concentrates the radiation by means of a parabolic mirror, 2 meters high and with an aperture of 1.2 meters, made of sheet zinc. By reflecting the wave from a conducting surface, he observes four distinct nodal points, and he finds the wavelength to be 66 centimetres. Hertz detects the sparking effect up to a distance of 16 meters, using a door aperture in the wall. He verifies the rectilinear propagation by means of the usual screening process; he demonstrates polarisation with metal gratings; he detects reflected rays at forty-five degrees; and he observes refraction through a huge prism "made of so-called hard pitch".

He comments at the end of his paper:

We have applied the term ray of electric force to the phenomena which we have investigated. We may perhaps further designate them as rays of light of very great wavelength. The experiments described appear to me, at any rate, eminently adopted to remove any doubt as to the identity of light, radiant heat, and electromagnetic ["elektrodynamischer"] wave motion.[62]

Hertz views propagation here under its 'optical" aspect, opening up the new field of the optics of the electromagnetic spectrum, which was rapidly developed by his successors.

Hertz's December 1888 paper is commonly accepted as representing the climax of his experimental work. Following its communication to the Berlin Academy and its publication in the *Annalen der Physik*, the fame of Hertz's work spread to an international audience. The modern reader of the paper is still struck by in Hertz's words the "demonstrative power" of the experiments. They have the virtues of simplicity, straightforwardness, and clarity. This demonstrative power was what Hertz had sought since he first began thinking about free propagation. In fact, in March 1888 he had attempted to "exhibit the propagation of induction through the air byw.ave motion in a visible and almost tangible form." In fact, he had then argued that in his previous experiment (paper no. 7), the first on the propagation in wires and in air,

though the inferences upon which that proof rested appear to me perfectly valid ... they are deduced in a complicated manner from complicated facts, and perhaps for this reason will not quite carry conviction to anyone who is not already prepossessed in favour of the views therein adopted.[63]

Aware of this defect in his first experiment, Hertz now, in March 1888, emphasised the phenomenological aspect and even stressed its quasi-independence from the theoretical interpretation. He maintained:

the demonstrative power of the experiment is independent of any particular theory. Nevertheless, it is clear that the experiments amount to so many reasons in favour of that theory of electromagnetic phenomena which was just developed by Maxwell from Faraday's views.[64]

The relation between theory and experiment that Hertz hints at in the latter passage is developed and made more precise in his two subsequent theoretical papers on electrodynamics in 1890 and in his introduction to his collected papers.

His "Fundamental Equations of Electromagnetics for Bodies at Rest", (paper no. 13) was one of the two theoretical papers with which he concluded his experiments on waves, rejecting the Helmholtian theory of the splitting of the electric force:

the splitting of the electric force into an electrostatic and electromagnetic part does not in these general problems convey any physical meaning which can be clearly conceived, nor is it of any great mathematical use; so that, instead of following earlier methods of treatment, it will be expedient to avoid it.[65]

6.8 An Analysis of Hertz's Logic of Discovery

In the foregoing pages I have given evidence for what I consider to be continuity in the development of Hertz's ideas from his March 1888 research (paper no.8) to his 1890 theoretical work (paper no. 13).

In this section, I analyse Hertz's rethinking of the entire process of his discovery of electromagnetic waves presented in his introduction to *Electric Waves*, the volume of his collected papers on electrodynamics. My aim is to underline that his reconstruction of his discoveries in the 1891 introduction is consistent on the whole with the development of his research as presented in his *Electric Waves*, the second volume of his *Gesammelte Werke*.

In the initial passages of the introduction Hertz speaks of Helmholtz's 1870 theory as a modification of the "standpoint of the potential theory", and he says it was deduced "from the older views [potential theories]" and from three hypotheses:

1   Changes of dielectric polarisations in [ponderable] non-conductors produce the same electromagnetic [electrodynamic] forces as do the currents which are equivalent to them.

2   Electromagnetic [electrodynamic] forces as well as electrostatic are able to produce dielectric polarisation.

3   In all these respects air and empty space behave like all other dielectrics.

In hypotheses I above, the adjective "ponderable" was skipped by the translator of the German original, although I deem that Hertz uses it to mean that his new view of air polarization does not belong to the older views. Hertz affirms that Helmholtz had set the two parts of the Berlin Academy problem on the grounds that the combination of these hypotheses with the electromagnetic laws, universally accepted in 1879, would yield Maxwell's equations.

The third hypothesis played an important role in Hertz's 1879 response to the Berlin Academy problem. In fact, the Academy problem required, according to Hertz,[68] an experimental test of either one or the other of the first two hypotheses above. The third one had been left aside, because to test all three seemed an unreasonable demand.

Having successfully verified the first hypothesis in 1887 (paper no. 6, discussed above) Hertz felt, nonetheless, that the third contained the main point of interest in Maxwell's theory, and that the first two hypotheses "would be proved simultaneously if one could succeed in demonstrating in air a finite rate of propagation and waves".[69] (for him, air ("Luftraum") and empty space ("leerer Raum"), i.e. ether, were synonymous).

Expressing in modern language Hertz's ideas, one can say that at this point Hertz understood that the core of Maxwell's theory was not polarisation current in material dielectrics but ether polarisation current in our terms, vacuum displacement current. Whereas in Helmholtz's theory the third hypothesis was introduced as a tentative extrapolation of polarisation effects from material dielectrics to ether, in Hertz's view ether polarisation assumed logical priority. He recognized that the polarisability of material dielectrics was marginal to Maxwell's displacement theory. This recognition represented the main aspect in Hertz's overturning of Helmholtz's electrodynamics and it was Hertz's way of understanding Maxwell's theory.[70]

It is noteworthy to remark that in his 1890 theoretical paper,[71] Hertz considered the "hypothetical polarisation of the dielectric ether" as Maxwell's own representation of electromagnetic phenomena. His reconstruction of his discoveries in the 1891 introduction is thus consistent with the development of his research as presente d in his *Electric Waves*.. Given these premises, it turns out that the study of Helmholtz's and Hertz's polarisation theories represent a prerequisite for an analysis of Hertz's logic of discovery in the process which led him to the momentous discovery of the "waves of electric force". Hertz's discussion of polarisation theory (PT), in the theoretical section of his introduction to the collections of his experimental and theoretical papers on electromagnetism, is qualitative only and, as such, needs to be combined with a study of Hertz's[72] and Helmholtz's[73] PT in their scientific essays. A short summary of PT from these essays is therefore presented in the following pages.

As shown above, Hertz's PT in 1888 reverses Helmholtz's PT because in Hertz's theory ether polarisation is a primitive (independent) concept, and it is by far the predominant effect in ether and in air. In fact, in material dielectrics the Poisson-type polarisation contributes to the predominant ethereal polarisation as a merely multiplicative factor.[74] Helmholtz's PT is a Poisson-type theory in the sense that polarisation of material dielectrics is an effect of induction acting-at-a-distance on bound charges enclosed within insulators. As a consequence of the finite rise-time effect in the motion of induced charges, a polarisation wave propagates with a finite velocity in insulators. The same type of PT can be extended to ether and, in this case, it predicts the possibility of longitudinal and transversal ethereal waves propagating with finite velocity (see section 5 A above). This PT is not however, in general, a contiguous theory, because the primary electric force acts at-a-distance in discontinuous material dielectrics and in ether. In Hertz's so called limiting-case of Helmholtz's PT,[75] action-at-a-distance disappears. However this case is considered by Hertz in 1891 as an artificial theory, not free from contradictions.[76] The predicted (by Helmholtz's theory) ethereal waves have a feeble intensity because ether has a small polarisation constant in respect to material dielectrics (i.e., it behaves very feebly as a Poisson-type dielectric).

In Helmholtz's and Hertz's PT, the concepts above are represented through a different interpretation of the polarisation constants.[77] In fact, in Helmholtz, air has a very feeble dielectric constant and waves in air should be so feeble as to be almost undetectable.[78] For this reason, Hertz, when he still accepted the Helmholtian standpoint, was preparing blocks of paraffin, a substance with a relatively high polarisation constant with respect to air. The detection of waves in air, clearly reported in Hertz's 1888 (paper no.8) thus supports Hertz's and discredits Helmholtz's PT.

Coming now to the problem of the role played by Hertz's PT in his discovery of the waves, I argue that Hertz, as early as the end of January 1888, began to interpret his experiments in terms of his PT and not Helmholtz's Poisson-type theory. Evidence concerning this point is found in Hertz's initial passage[79] in his 21 January, 1888 essay no.7:

If variable electric forces act in the interior of insulators whose dielectric constant differs appreciably from unity, the polarisations which correspond to those forces sexert electrodynamic effects [according to Helmholtz' theory]. But it is quite another question [i.e. it is a different effect from the one predicted by Helmholtz' theory] if in air as well variable electric forces are connected with polarizations of electrodynamic action. We may conclude that, if this question is to be answered in the affirmative, electrodynamic actions must be propagated [in air] with a finite velocity [the translation from the German and italics are mine].[80]

Let us call this statement P(1). In P(1) Hertz differentiated his PT from Helmholtz's. In fact, Helmholtz's PT predicted[81] finite propagation of currents in dielectrics (*Isolatoren*) whose dielectric constant differs appreciably from one, a propagation which somehow extended to dielectrics that propagation in conductors (wires) known by theory and experiment since Weber, Kirchhoff, von Bezold, and Kelvin. But, as Hertz stated above, propagation in air is quite another question: waves in air of detectable intensity are predicted only by Hertz's and not by Helmholtz's PT.

In short, in 1888 Hertz, in order to differentiate his own theory from Helmholtz's position, stressed that the novelty of his discovery involved: the existence of waves in air (Luftraum) or equivalently in ether or a vacuum (Leherraum).[82] He distinguished the theoretical interpretation of waves in air from the interpretation of waves in material dielectrics.

Hertz again took up the same argument presented in P(1) in 1891 in his introduction to *Electric Waves* . There he dealt with a rational reconstruction of his dramatic and sudden paradigm shift in 1888 from Helmholtz's PT to his new PT. He argued that the main point of interest in his new theory was his interpretation of the third hypothesis of the Academy prize.

I repeat here the Academy's third hypothesis with some elucidation:[83]

In all these respects [i.e. as regards the interaction between electrostatic and electromagnetic forces and variable polarizations, a consequence of the combined first and second hypotheses] air and empty space [ether] behave like all other dielectrics(parentyheses[] are mine).

Let us call this P(2). P(1), Hertz's statement of his new 1888 conception, and P(2), the Academy third hypothesis, contradict each others because in P(2) ether polarisation is likely to appear as an extension of material dielectrics polarisation, while in P(1) the two polarisations are different. In fact, P(2) expresses the point of view of Helmholtz's PT, while P(1) is Hertz's PT.

This interpretation is confirmed if one takes into account the omission[84] of a sentence in the English translation of Hertz's original passage in his introduction which I mentioned above. In the omitted sentence,[85] Hertz commented that waves in a given insulator (*bestimmte Isolator*), a consequence of the combined two first hypotheses, are no surprise (*nicht sehr überraschen*), no more than the long known phenomenon of propagation with a finite velocity of the electric excitation (*Erregung*) in wires. When mentioning waves in a given insulator, Hertz evidently hinted at Helmholtz' theory which, consistent with P(2) above, predicted polarisation waves in dielectrics. It is here that the polarisation theme interweaves with the wire-waves propagation theme.

Hertz suddenly discovered that P(2) was not Maxwell's theory and that in his Maxwell-type theory the third hypothesis was not to be interpreted as a consequence of the first two. In short, Hertz's turning point consisted in his original and anti-Helmholtian interpretation of the third hypothesis. In fact, he clearly stated:
"I felt that the third hypothesis contained the gist and special significance [*der Kernpunkt*] of Faraday's, and therefore of Maxwell's, view, and that it would thus be a more worthy goal for me to aim at".[86]

In clear words, Hertz discovered that the main point of interest in the new theory did not reside in the interpretation of the third hypothesis, as meaning that polarisation in empty space and air are secondary effects *à la* Poisson (a consequence of the other two hypotheses), similar to the well 'known insulators' polarisation. This type of polarisation would explain the long known waves in insulators and in wires. But the waves he detected in air are another thing. These types of waves are predicted only by a Maxwell-type theory, not by Helmholtz's.

In conclusion, as regards the wave propagation theme, above I have compared Hertz's passage in his 1888 paper (no. 7) with his passage in his 1891 introduction to *Electric Waves*. In both passages Hertz remarks that detecting waves in material dielectrics or waves of current in a wire is irrelevant for an experimental test of Maxwell's theory.

The identity of points of view in these two works by Hertz represent another point supporting the thesis that Hertz's introduction is a reasonably faithful reinterpretation of the logic of research that guided him in his experiments.[87]

CHAPTER 7

HERTZ'S 1884 THEORETICAL DISCOVERY OF ELECTRO-MAGNETIC WAVES

7.1. Hertz's 1884 Principles of Uniqueness and Independent Existence of Electric and Magnetic Forces and his 1888 Discovery of Electromagnetic Waves

In his early research on electrodynamics, published in 1884 with the title *On the Relations between Maxwell's Fundamental Electromagnetic Equations and the Fundamental Equations of the Opposing Electromagnetics*,[1] Hertz advanced an alternative conception of the electric field:

According to Faraday's idea, the electric field exists in space independently of and without reference to the methods of its production; whatever therefore be the cause which has produced an electric field, the actions which the field produces are always the same.[2]

Faraday's idea was an alternative to the action-at-a-distance view of forces then predominant in Germany. Starting with Faraday's view that the electric field exists in space independently from its source, Hertz then concluded that there is a unique electric and a unique magnetic force.[3] He raised the above conception to the role of a *principle of uniqueness* (*Einheit*) of electric and magnetic forces:

This principle is the necessary presupposition and conclusion of the chief notions which we have formed in general of electromagnetic phenomena.[4]

He related this principle of uniqueness to Faraday's conception of an independent existence of forces in space, considering the former to be a necessary presupposition for the latter.[5] He intended the uniqueness principle to be a founding principle of his electrodynamics. By *independent existence* he clearly meant "independent from their sources". In fact static and dynamic electric forces have different sources, *i.e.*, charges, conduction currents and displacement currents, hence their sameness (identity) can be related to their uniqueness only if the causal correlation between force and source of force is broken and their independent existence is affirmed.[6]

The principle of uniqueness was used by Hertz in 1884 to derive Maxwell's equations from Neumann's (pre-Maxwellian) electrodynamics. In Hertz's opinion it represented the missing link that served to derive "Maxwell's fundamental equations from the fundamental equations of the opposing electrodynamics".

In his essay Hertz developed a procedure for obtaining Maxwell's equations from continental electrodynamics. He set out to show that the vector potential for steady or slowly varying currents, on which the old electrodynamics had relied exclusively, are incomplete, and he calculated by an iterative process the missing parts. Once he had derived the complete potentials or, as they were later called, the "retarded potentials", Hertz proceeded to his main concern: Maxwell's theory.

Hertz's premise was that Ampere had predicted the existence of ponderomotive forces between electric currents as a consequence of the identity of the ponderomotive forces exerted by a magnetic pole with those exerted by an electric current. This identity was, in Hertz's opinion, an assertion of the "unity" of the magnetic force. He formulated an analogous principle for the electric force:

Those electric forces which have their origin in inductive actions are in every way equivalent to equal and equally directed forces from an electrostatic source.[7]

He regarded the "principle of unity of electric force as a necessary presupposition and conclusion of the chief notions which we have formed in general of electromagnetic phenomena".[8]

From the two principles of unity together with the accepted laws of electric and magnetic actions of closed currents, and from principles such as those of energy conservation, action and reaction, and the superposition of forces, Hertz deduced a new electrodynamics. His theory was valid only for closed circuits, unlike Helmholtz's which was also valid for open circuits.

Hertz argued as follows. If there is a unity of electric force, a magnet of varying intensity should set in motion a charged body with the same force with which it induces a current. By the principle of action and reaction the charged body should in turn set the magnet in motion. Further, two magnets of varying intensities should attract or repel each other with forces depending on the time rate of variation of the magnetic force. Such magnetic actions are not only omitted from but are also in contradiction with the laws governing the constant forces of the old electrodynamics.

From energy conservation and the unity of magnetic force, the correction in the magnetic actions leads to a correction in the induced electric forces. This correction in turn requires a second correction in the magnetic forces because of the essential unity of forces between currents and forces between magnets. One obtains an infinite series of successive approximations.

In this way, Hertz begins by defining a magnetic current as the rate of change in magnetic polarisation of a "ring-magnet" (toroidal magnet) of variable magnetic intensity. Due to the unity of electric force, the "potential of the ring-magnet on an electric pole can, apart from its multiplicity, be represented by the potential of the double layer on the pole".[9]

Using an analogy with Ampere's theorem for magnetic double layers, he assimilates the ring-magnet to an electric double layer bounded by a magnetic current. Hertz's theory is symmetric in electric and magnetic forces and currents, since magnetic currents interact according to the same laws as electric currents.[10] As a consequence of this symmetry, he obtains a new, magneto-electric induction law:

> Two ring-magnets which are placed close together and side by side will attract each other at the moment when they both lose their magnetism if they are magnetised in the same direction; they will repel each other if oppositely magnetised.[11]

The interaction of ring-magnets is a new type of action that is missing, according to Hertz, in the "ordinary electrodynamics" exemplified by Franz Neumann's work. Another "new" effect that he derives is the alternation of the magnetic polarisation of a ring-magnet when it is rotated around an axis perpendicular to an electrostatic force. Yet another new effect is the motion of electric charged bodies produced by a ring-magnet of diminishing intensity.

The new magnetic forces affect, in their turn, the electromotive forces and, therefore, the electric induction forces; this interaction of forces is evident from the Helmholtian procedure of deriving induction forces from ponderomotive magnetic forces between circuits. But this, in turn, will entail a further modification in the force of magnetic interaction, and the argument will be repeated.[12]

Hertz developed a mathematical treatment for these qualitative ideas. He introduced an electric vector potential $\mathbf{U}$ in the usual way:[13]

$$\mathbf{U} = \int \frac{\mathbf{u}}{r} d\tau \quad \text{div} = 0$$

where r is the position vector, dt the element of volume and **u** the current density. The expression is valid also for the case of a current of variable density. Hertz defines the first order magnetic field $\mathbf{L_1}$:

$$\mathbf{L_1} = -A \, \text{rot} \, \mathbf{U}_1 \quad , \quad (1)$$

where A is Weber's conversion factor and the reciprocal of the velocity of light in a vacuum. Due to the principle of conservation of energy, variations of U_1 produce a first order electric force X_1:

$$\mathbf{X}_1 = -A^2 \frac{d\mathbf{U}_1}{dt} \quad . \quad (2)$$

On the basis of Hertz's premises, electric forces $\mathbf{X_1}$ must appear even if the magnetic forces $\mathbf{L_1}$ have their origin in an arbitrary system of variable magnets; i.e., in magnetic currents. He then develops the mathematics of magnetic currents by introducing a magnetic vector potential $\mathbf{P}_1$[14]

$$\mathbf{P}_1 = \int \frac{\mathbf{p}}{r} d\tau \, , \quad \mathbf{p} = \frac{d\lambda}{dt} \, ,$$

where l is the "magnetic polarisation" and **p** the "magnetic current." He relates the electric forces $\mathbf{X}_1$ to the magnetic vector potential:

$$\mathbf{X}_1 = A \, \text{rot} \, \mathbf{P}_1 \quad (3).$$

The same considerations that led from the potential of electric currents to the inductive forces (2) allow one to infer from (3) the existence of induced magnetic forces L_1:

$$\mathbf{L}_1 = -A^2 \frac{d\mathbf{P}_1}{dt} \quad . \quad (4)$$

These equations, like equations (1) and (2) are valid for both electric and magnetic currents. Hence electric forces (2) can be represented in form (3). But if they are variable they generate magnetic forces of form (4).

In accordance with the qualitative argument above, however,

the existence of forces in equation (4) entails a correction at the forces in equation (1). To calculate this correction, Hertz first solves the general problem of expressing the force $\mathbf{X} = -A^2 \dfrac{dU}{dt}$ in the form $\mathbf{X} = A \operatorname{rot} \mathbf{P}$, i.e., of expressing forces (2) in form (3). He does this by equating the expressions for $\mathbf{X}_1$ in equations (2) and (3):

$$-A^2 \frac{dU_1}{dt} = A \operatorname{rot} \mathbf{P}^*, \quad \operatorname{di} \mathbf{P}^* = 0.$$

Hence,

$$\nabla \mathbf{P}^* = A \frac{d}{dt} \operatorname{rot} \mathbf{U}_1 \quad ,$$

and

$$\mathbf{P}^* = -\frac{A}{4\pi} \frac{d}{dt} \int - \frac{\operatorname{rot} \mathbf{U}_1}{r} d\tau \quad ,$$

$\mathbf{P}^*$ now represents a correction to P_1; consequently L_1 must be corrected in equations (4) and (1). The correction is:

$$-A^2 \frac{d\mathbf{P}^*}{dt} = -\frac{1}{4\pi} A^2 \frac{d^2}{dt^2} \int \frac{\operatorname{rot} \mathbf{U}_1}{r} d\tau \quad ,$$

It is evident that the correction can be expressed in both forms (4) and (1) because of the possibility of interchanging space and time operators. Let us call $\mathbf{L}_2$ the corrected magnetic force in the form of equation (4):

$$L_1 = -A^2 \frac{d(\mathbf{P} + \mathbf{P}^*)}{dt} = -A^2 \frac{d}{dt}\left(\mathbf{P}_1 + \frac{A}{4\pi}\frac{d}{dt}\int -\frac{\operatorname{rot} \mathbf{U}_1}{r} d\tau\right) ,$$

The corrected equation (1) is:

$$L_2 = L_1 + -A^2 \frac{d\ P^*}{dt} = -A\ \text{rot}\ (U_1 + U^*) = -A\ \text{rot}\ U_2 \quad .$$

Hence:

$$U_2 = U_1 - \frac{1}{4\pi} A^2 \int \frac{U_1}{r}\ d\tau \quad .$$

Through the same argument that derives U_2 from U_1 and U^*, the electric vector potentials of higher order U_3, U_4, ..., can be found:

$$U_3 = U_2 - \frac{1}{4\pi} A^2 \int \frac{U_2}{r}\ d\tau \ ,\ etc.$$

The reiterative process is mathematically equivalent to a series summation. The series converges, according to Hertz, towards the actual electric and magnetic vector potentials U, P. He demonstrates the convergence to be a special case of sinusoidal variability.[15] The potentials U and P in empty space appear to obey d'Alembert's equations and are propagated with a velocity equal to the reciprocal of Weber's conversion coefficient i.e., the velocity of light in a vacuum:

$$\nabla U - A^2 \frac{d^2 U}{dt^2} = 0 \qquad \text{div}\ U = 0 \quad ,$$

$$\nabla P - A^2 \frac{d^2 P}{dt^2} = 0 \qquad \text{div}\ P = 0 \quad .$$

Although, as Hertz noted, Riemann in 1858 and Lorenz in 1867 had derived equations similar to the above, they had done so by different routes. Neither had regarded perturbative effects of higher order terms as consistent with general principles and as experimentally detectable. Hertz defined the "completely corrected forces" **X** and **L** as:[16]

$$\mathbf{X} = -A^2 \frac{d\mathbf{U}}{dt} \quad \mathbf{L} = -A \ \text{rot} \ \mathbf{U} \ .$$

He then eliminated the potential functions **U** and **P**:

$$A \frac{d\mathbf{X}}{dt} = \text{rot} \ \mathbf{L}, \quad \text{div} \ \mathbf{X} = 0,$$

$$A \frac{d\mathbf{L}}{dt} = \text{rot} \ \mathbf{X}, \quad \text{div} \ \mathbf{L} = 0 \ .$$

Here for the first time the equations for the electric and magnetic forces are written in a symmetric form. Hertz comments at this stage:

The system of forces given by these equations is Maxwell's. Maxwell found them by considering the ether a dielectric, in which a changing polarisation produces the same effects as an electric current. We have reached them by other premises, generally accepted even by the opponents of the Faraday-Maxwell view.[17]

Hertz deduced the equations by an alternative route from Maxwell's. He emphasised the generality of his route, having avoided Maxwell's special assumption of a dielectric ether. He regarded Maxwell's system together with its deduction "as the most obvious from a certain point of view", but at the same time as not "necessary". Hertz's derivation of Maxwell's system from the old electrodynamics exposed the inconsistencies of the latter. He was then confronted with the dilemma of admitting that a necessary truth can be a necessary consequence of a false premise. He concluded that other systems besides Maxwell's were possible and that they could be as exact as Maxwell's. One was certainly Helmholtz's.

Hertz regarded Maxwell's theory as more "complete" relative to "the usual system of electrodynamics" and the simplest in respect to other possible theories.[18] Hertz does not claim that his derivation has the character of a logical demonstration, recognising that Maxwell's equations are not the only possible modification of the classical ones. He argued, however, that "Maxwell's theory has the advantage that it does not contain within itself the proof of its own incompleteness". Furthermore, it supplies a simpler way than other possible theories of representing electrodynamic phenomena. In any case, the fact that it provides a theory for such basic features as the attraction between variable magnetic shells makes it superior to others.

The two principles of the unity of the electric and magnetic forces, which play a central role in the derivation of the equations representing Maxwell's forces, are presented by Hertz as being somehow related both to Faraday's views and, though less markedly, to Weber's elementary law. Since this dual indebtedness is relevant to the future development of Hertz's thought, I will quote at length his gloss on the principle of the unity of electric force in his 1884 essay.

This principle is the necessary presupposition and conclusion of the chief notions which we have formed in general of electromagnetic phenomena. According to Faraday's idea, the electric field exists in space independently of and without reference to the methods of its production whatever therefore be the cause which has produced an electric field, the actions which the field produces are always the same. On the other hand, by those physicists who favour Weber's and similar views, electrostatic and electromagnetic actions are represented as special cases of one and the same action-at-a-distance emanating from electric particles. The statement that these forces are special cases of a more general force would be without meaning if we admitted that they could differ otherwise than in direction and magnitude, that is, in the nature and mode of action.[19]

In 1884, Hertz also cites Weber's elementary law in support of the unity of force in the sense that this unique law includes forces of the two types. His citation is part of his strategy of presenting the uniqueness principle as comprehensive, applying to the otherwise contrasting positions of Faraday and Weber. It is a sign of the importance he attributed to the principle at this stage.

In 1888, when Hertz succeeded in demonstrating the propagation of electric force in air and shortly before his interpretation of radiation in the Hertz-Maxwell context of a unique force, he stated the independent existence of the force in almost the same words as in 1884:

The most direct conclusion [from these experiments] is the confirmation of Faraday's view, according to which the electric forces are polarisations existing independently in space. For in the phenomena which we have investigated such forces persist in space even after the causes which have given rise to them have disappeared.[20]

Hertz's 1884 article, which he published in the *Annalen der Physik*, found some reception in German scientific circles. It was not well received, however, by Helmholtz's pupils, as is shown by the debate on the article prompted by a series of articles in the *Annalen* by E. Aulinger and L. Lorberg. Aulinger in 1886, following a suggestion by Boltzmann, finds an inconsistency in Hertz's formulation. Aulinger emphasises that forces originating from currents are composed of electrostatic and electrodynamic parts, and prefers to formulate the principle of the unity of electrical forces as follows: "Once the forces which act on a static or on a uniformly moving electrical charge are determined, the whole of the electrical force is determined".[21] He argues that from this unique principle all of Hertz's conclusions follow. Whereas Hertz invoked Weber's theory in defence of his principle, Aulinger proves that Weber's theory is contradicted by the principle of the uniqueness of force in Hertz's form. Aulinger affirms that it is not his intention to defend Weber's theory; rather, he believes that Hertz's principle, in his own formulation, has a great deal of plausibility.

Following an intervention by Lorberg in the 1886 *Annalen*, Boltzmann entered the debate in a conciliatory fashion. Boltzmann proposed not to discuss the "aprioristic" probability of the statements any further, but to discuss experiments. He pointed out that the action of a changing magnetic field on an unmoving static charge could be tested experimentally, settling the question once and for all. This experiment would represent as well an "experimentum crucis" for Weber's electrodynamic theory.[22]

The same volume of the 1887 *Annalen* that contains Lorberg's reply to Boltzmann contains Hertz's report on his initial experiments dating from his Karlsruhe period, "On Very Rapid Electric Oscillations," followed in quick succession by two other articles on the same subject. Hertz's interest in the theoretical debate was superseded by the new challenging experimental approach to rapid oscillations, a matter which he felt was somehow connected to the same problem of deciding between the two contrasting electrodynamics.

At this point I want to outline briefly the main conceptual features of Hertz's 1884 article:

a) Hertz sees the propagation of electrical and magnetic forces as the result of the structure of electrical and magnetic forces and potentials as expressed in the equations, a characteristic shared by the continental theories of retarded action of Riemann and Lorenz. He seeks probative evidence for the propagation of electric and magnetic forces in electromagnetic phenomena, and not, as Maxwell did, in optical phenomena.

b) In 1884 Hertz pays no attention to physical hypotheses about a medium as the supporter of fields in the fashion of Faraday and Maxwell; in contrast the conception of a polarisable ether is the main pillar of his 1888 experiments and 1890 theory. In 1884, Hertz's methodology is formalistic in that he proceeds by generalising from Franz Neumann's theory.

c) In 1884 Hertz deduces Maxwell's equations from fundamental principles, without regarding the deduction as a rigorous proof that Maxwell's system is the only possible one; contrastingly, in 1890 he does not deduce Maxwell's equations from any prior principles, but postulates them and places them at the head of his theory. His exclusion in 1884 of any claim to theoretical uniqueness is one aspect of Hertz's epistemology, which he fully developed later in his *Principles of Mechanics* (See below in this book).

d) The independence of fields from sources is not contrasted with Helmholtz's theory in 1884, but a few years later it is; then it plays a central role in Hertz's rejection of Helmholtz's duality of electric forces.

e) The 1884 theory was a mathematical theory of contiguous propagation of waves of electric and magnetic fields in space, which Hertz achieved through a clever combination of the reciprocal interactions between Ampére's and Faraday's forces without any interpretation of the fields as ether polarisations; rather he considered it a merit that he derived Maxwell's equations without a physical hypothesis about ether.[23] In so doing in 1884, he differentiated the idea of contiguous propagation from that of a substantial ether.

f) Hertz did not claim that his 1884 derivation had the character of a demonstration, i.e., of a logical deduction from premises.[24] He rather presented it as a possible modification of the traditional theory, thus admitting that other modifications besides Maxwell's were possible; one example was certainly provided by Helmholtz's work. However, he regarded Maxwell's system "as the most obvious from a certain point of view".[25] Hertz's 1884 ideas concerning the relation of his theory to other electrodynamical theories are remarkably consistent with his later conceptions on the plurality of theories and with the ideas adopted in his *Prinzipien*.

I argue that in 1888 Hertz thought that his discovery of waves of electric force was consistent only with his 1884 theory and not with Helmholtz's position.[26] My evidence for this thesis is that in his "On the Finite Velocity of Propagation of Electromagnetic Actions", dated 2 February 1888, when Hertz succeeded in interpreting his experimental results as a contiguous propagation of electrical forces in air, he revived his 1884 conception of independent existence, using the same words as he used in 1884, just adding the new view of electric force as space polarisation, a view he also ascribed to Faraday:

> The most direct conclusion [from these experiments] is the confirmation of Faraday's view, according to which electric forces are polarisations existing independently in space. For in the phenomenon which we have investigated such forces persist in space even after the causes which have given rise to them have disappeared.[27]

The similarity of the above quoted 1884 and 1888 reports is evidence of the determinant role played by the independent existence concept in effecting Hertz's conceptual shift to the contiguous propagation view. In the last lines, the conclusion that forces persist in space even after the disappearance of the cause which generated them, i.e., their sources, is clearly a reaffirmation and an extension of the 1884 concept of their independent existence.

It can be concluded that in 1888, when Hertz radically changed his standpoint, he understood that this new position was consistent with his 1884 conception. Moreover, in a later (19 March 1890) theoretical paper "On the Fundamental Equations of Electrodynamics for Bodies at Rest",[28] in which Hertz presented his own version of Maxwell's theory, he reaffirmed[29] his 1884 conception of uniqueness and he concluded that there is a *unique type of electrical force in radiation*, a conclusion which cleared up for him the otherwise embarrassing behaviour of electrical radiation in his experiment.[30] In the same paper he assumed[31] that the *state of polarisation of ether* was an independent (primitive) concept (i.e., not in need of any further reduction), thus reversing [32] the ideas of traditional and Helmholtian Poisson-type polarisation, according to which static and dynamic polarisation are an effect of charge induction.

In the same paper Hertz introduced[33] the "velocity of propagation of electric and magnetic oscillations as an intrinsic constant (*innere Constante*)" of the ether and the specific inductive capacity (*Dielektricitätsconstante*) and the magnetic permeability (*Magnetisierungsconstante*) as extrinsic (not intrinsic) constants of a substance, just as he had done in his 1884 paper.[34]

7.2 Why did Hertz not mention his 1884 Paper in 1888?

For an appraisal of the influence of Hertz's 1884 article on the later development of his ideas, it is important to note that he never explicitly mentioned it in his later articles. In particular, he never returned to that peculiar derivation of Maxwell's equations which, as we saw, he had worked out in his earlier contribution.

This indifference seems at first surprising in the light of the following detail: the symmetric form in which Hertz writes the equations for Maxwell's electric and magnetic forces in 1884 is exactly the same as the one he writes in 1890 in his article, "On the Fundamental Equations of Electrodynamics for Bodies at Rest".[35]

In 1890, Hertz, stressing their axiomatic foundation, seems to have forgotten his prior symmetric rendering of the equations. In a passage there he even acknowledges Oliver Heaviside's priority on this matter:

> Mr. Oliver Heaviside has been working in the same direction ever since 1885. From Maxwell's equations he removes the same symbols [the vector potentials] as myself; and the simplest form which the equations thereby attain [in Heaviside's papers in 1885 and 1888] is essentially the same as that at which I arrive. In this respect, then, Mr. Heaviside has the priority.[36]

In the section on Hertz's experiments,[37] I posed the two questions why Hertz did not resume his 1884 ideas in 1887, when he, starting his research on rapid electric oscillations afresh, adopted instead Helmholtz's theory, and why he did not mention explicitly his 1884 symmetric deduction of Maxwell's equations again in 1888, when he shifted to a Maxwell-type theory.

I think that one must understand Hertz's apparent amnesia as part of his general attitude towards his 1884 paper. In 1887-1888 he began his research from a completely different perspective than in 1884. In this new approach, equations of the 1884 type were out of place; the 1884 Hertzian conception of the unity of force was in strident contrast with the Helmholtian principle, which Hertz supported in 1887, of the different nature of static and dynamic forces.[38] Hertz could not graft the unity principle onto Helmholtz's theory without destroying it. In 1887 he guided his new experiments by the secure framework of the more familiar and authoritative Helmholtian theory.

However, in 1888, though he did not mention his 1884 essay, he explicitly revived his 1884 conception of the independence of fields (Sections 6.5, 6.6, 6.7, 7.1) My explanation is the following: in 1888 he was led to a standpoint in which the unity of fields both in its primary meaning as the "unity of electric fields in radiation" and in its meaning as the independence of fields from sources became central to his theory of radiation.

I also find it understandable that in 1890 he did not connect his symmetric equations with the equally symmetric ones in his 1884 research. In 1890 his conception of an axiomatic foundation of theories as the purpose of physical inquiry (ideas expressed later on in his *Prinzipien*) placed him in a different context from that inherent in his 1884 derivation of Maxwell's equations from other premises.

This remarkable difference in methodological context helps to explain why Hertz did not recognise or mention explicitly his earlier introduction of the symmetric equations.

In order to support my answers to these questions, I also refer to my previous analysis of Hertz's and Helmholtz's PT. Helmholtz's PT had both a progressive and regressive aspect: the progressive aspect offered a strong physical conception of the Maxwellian concept of displacement in ether, by interpreting it as an extension of the easily observable Poisson-type polarisation of material dielectrics. On the other hand, the same displacement, in its regressive aspect, risked being interpreted in Helmholtz's fashion, as a mere extrapolation to ether of the effect of material polarisation, a fact of secondary conceptual relevance.

In 1884, Hertz had accepted contiguous propagation and derived Maxwell's equations without introducing dielectric action and the related PT,[39] founding his derivation on purely dynamical considerations of the effect of Faraday's induction on Ampere's action-at-a-distance forces, and adopting the principle of independent existence.

If one takes into account the above rather complex situation, some light is shed on the above questions:

1) At the start of his experimental research in 1887, lacking a physical conception of dielectric action, the principle above of independent existence, successfully used in 1884, appeared as a formal principle rather empty of physical content.

2) In his initial 1887 approach to his experiments, Hertz's strategy consisted in taking advantage of the progressive aspect of Helmholtz's PT thus overcoming the weakness in his 1884 theory: the need for a physical conception of polarisation. In 1887 equations of the 1884 type were out of place. Moreover, his 1884 conception of the unity of force was in strident contrast with the Helmholtian tenet, which Hertz still adopted[40] in November 1887, of the different nature of static and dynamic forces. Thus in 1887, Hertz began his research from an Helmholtian PT standpoint, stressing the role of dielectric Poisson-type polarisation, a view that in 1884 he had considered different from his own.[41]

3) In March 1888, Hertz soon succeeded in neutralising the regressive aspect of Helmholtz's PT by appealing to an innovative notion independent of his 1884 essay: ether polarisation, a primitive concept of his theory. This notion avoided the trap of a strict empirical interpretation of the polarisation concept, one from which the same Helmholtz did not escape, and one which guided Hertz's fol-

lowing experiments.

In conclusion, in 1887, Hertz guided his new experiments by the more secure support of the familiar and influential Helmholtian polarization theory. It is, therefore, understandable that he ignored his 1884 paper. In January 1888, he appealed to an innovative interpretation of the "independent existence" of his 1884 essay, by conceiving ether polarization as a primitive concept of his theory. The radicalism of which interpretation might explain why in 1888 he considered his 1884 way of introducing the unity principle by the independ existence to be outmoded and he did not mention explicitly his 1884 work, although he reproduced almost entirely some of its passages.

Edmund Hoppe, who taught at Gottingen at the turn of the century, maintains that in 1884 "Hertz had already recognised the basic importance of his two equations, and one can explain the fact that he did not begin [his theory] with them, as Heaviside did, as the result of the situation at that time in Germany".[42] Hoppe says that Hertz in 1884 had a "pedagogical" interest in establishing a connection between the old and the new electrodynamics. Hoppe seems to ignore the priority of Hertz's 1884 symmetric formulation of the equations, and, like Hertz himself, stresses Heaviside's priority in the axiomatic foundation of Maxwell's theory. Inasmuch as I accept that there was a difference in the situations in 1884 and 1890, I agree with one of Hoppe's theses. I argue that Hoppe's mentioning of a situation that prevented Hertz from adopting from the start in 1887 his early 1884 conception, is to be intended as his reference to the cultural and institutional indebtedness of the young physicist Hertz to his celebrated master Hermann von Helmholtz. Hoppe adds that Hertz in 1884 had a "pedagogical" interest in establishing a connection between the old and the new electrodynamics.[43] While I do not exclude that the reasons above could have motivated Hertz's position, I believe that conceptual reasons of the type I presented above also had a determinant role.

In contrast with my theses, Hertz's 1884 contribution is considered by Buchwald[44] as a failed attempt at a conciliation between Maxwell's and Hertz's essentially different theories. I think I found adequate evidence for my thesis above in Hertz's original contribution to electrodynamics and in many passages reported in his papers. Another piece of evidence is presented by Max Planck's authoritative comments (See forward to part two).

One might be surprised by the complexity of the panorama of Hertz's discovery, but it should be remembered that a great discovery is always a complex affair.

CHAPTER 8

A FOUNDATION FOR THEORETICAL PHYSICS IN HERTZ'S INTRODUCTION TO *DIE PRINZIPIEN DER MECHANIK*

8.1. Hertz's Bild-*Conception*

In reading Hertz's technical papers on his experiments in Electrodynamics, his reflections on them in his theoretical papers[1] and his *Bild* conception of physical theory in his most conclusive work[2], one is led to see a thread which, more or less evident at the beginning and more clearly apparent at the end, interweaves all of Hertz's thought. (One can also use the term *intersections*, to indicate the intricacy of themes in this thought.) These interweavings are also manifest on a mere philological level, through the recurrence of certain words and sentences in his writings. The interweavings in Hertz's works are highlighted in my previous essay on the role played by his 1884 conceptions in his 1888 discovery of contiguous action and electromagnetic waves.

Another example thereof is offered by Hertz's *Bild* conception of theory in his *Prinzipien*,[3] which is largely the result of his reflections on his experiments and research in electrodynamics.

This being the case, it is the more surprising that, to my knowledge at least, only a few papers and those exploratory, have been devoted to a comprehensive enquiry into the whole of Hertz's work.[4] I am convinced that such an inquiry would lay the ground work for a better understanding of Hertz's contributions to physics.

To begin with, I will summarise briefly Hertz's presentation of his *Bild* conception of theories in his *Prinzipien*.

For Hertz, physical theories are *Bilder* (or *Scheinbilder*), representations (not descriptions) in our mind of the external world. According to Hertz, the value of *Bilder* as scientific theories depends on their meeting three requirements: Permissibility *(Zulässigkeit)*, Appropriateness *(Zwegmässigkeit) and* Correctness *(Richtigkeit)* (Henceforth, I will abbreviate these words PM, AP and CR, respectively).

As regards PM, Hertz's own comment is that *Bilder* must be "logically permissible - or, briefly, they should be permissible". He explains:

those *Bilder* are inadmissible [*unzulässig*] which explicitly contradict the laws of our thought" [*die Gesetze unseres Denkens*].[5]

The "customary representation" (*Darstellung*) of Mechanics, is, for Hertz, defective as regards PM, although it satisfies the demand for CR. In fact the concept of action and reaction as applied to a circular motion and, in general, the relationship between external and internal (inertial) forces are logically unclear:

Can we, without destroying the clearness of our conceptions, take the effect of inertia twice in account, firstly as mass, secondly as force?

He then criticises the various formulations of the mechanics of Newton, Laplace, Lagrange, Thomson and Tait because they show a marked difference in their ideas, a situation which "in a logically complete science... is utterly inconceivable".

The demand for PM is fully met in Hertz's own mechanics by excluding force as a primary concept and by adding "hidden masses", which compensate for the exclusion of forces, in describing natural motions in the form of inertial motions in a multidimensional space. In fact, as shown above, the concept of force is considered to be a repetition of that of mass, thus impairing the logical clarity of the whole system.

In Hertz's thought AP deals with that feature of theory which "pictures the essential relations of the object", and thus produces "more distinct and simpler" *Bilder*. It has to do with notations, definitions, abbreviations, etc.

Finally, CR concerns the relation between concepts and observables, "the results of experience". "Customary mechanics" (i.e. the Newtonian-Laplacian *Darstellung*) is acceptable as regards CR, (i.e. "all those characteristics of our image, which claim to represent observable relations of things") but, being "logically" indefinite, is defective as regards PM.

Hertz poses the question: does the requirement of PM add "arbitrarily" to the "essential and natural" characteristics of things, resulting in the loss of AP? One might even consider the introduction in customary mechanics, of the concept of force to be an initial achievement in PM and a loss in AP, because in many cases, e.g., in celestial motions, we do not observe forces; however, on reflection, the concept of force has proved to be more a concession to AP then to PM

> Even if the forces have only been introduced by ourselves into nature [Hertz means: for the sake of PM, against AP], we should not, on that account, regard their introduction as inappropriate.[6]

In other words, in the past, a theory which, seeking PM, has introduced the very useful concept of force has enhanced in the end its AP.

However, in Hertz's time, a theory without forces has proved to possess more PM, although apparently losing its AP.

If, for the sake of PM, the system "includes very many motions which are not natural", this can be accepted because phenomena could be discovered in the future, that will account for these seemingly unnatural motions (pace AP): the electromagnetic motions are, for Hertz, an example of the heuristic power (in our words) that mechanics derives from the introduction of "unnatural" motions.[7]

Although Hertz is aware of the possible danger of seeking PM at the cost of apparently renouncing AP, he places utmost value on the axiomatic structure of a system (its *Syntax*), to the point that now even AP might apparently be sacrificed to PM. Seeking PM, in fact, could produce:

> A system of principles [that] embraces all the natural motions... but it also includes very many motions which are not natural. A system which includes the latter, or even a part of them, would picture more of the actual relation of things to each other, and would therefore in this sense be more appropriate.

In fact Hertz believed that his "hidden masses", a concession to PM, offer a solution to the problem of the mechanical explanation of electrodynamics pursued by Maxwell.[8]

8.2. Hertz's Criticism of Energetism

Thus Hertz defined the mode of representation founded on the laws of transformation of energy (*Energetism*):

> [This representation]...likes to treat the phenomena which occur in its domain as transformation of energy into new forms and to regard as its ultimate aim the tracing back of phenomena to the laws of the transformation of energy....[It is] a second image of mechanical processes which is of much more recent origin than the first...system of mechanical principles.[9]

By a system of mechanical principles Hertz meant that system which adopts force, space, mass and time as primitive concepts. In the same passage he expresses the opinion that *energetism* as a general conception of physical theory was "influenced by the overpowering impression made by the discovery of the principle of conservation of energy".[10] It is clear that Hertz distinguishes here between the principle of conservation of energy and *energetism*.. While he accepts the former, using it on many occasions in his development of the theories of electrodynamics and mechanics, he dismisses energetism in favour of his *third image of mechanics,* i.e., a system of mechanics based only on the primitive concepts of mass, space and time.

As is known, energy conservation had been raised in 1847 to the status of a general principle in physics by Helmholtz, who showed how it could be successfully used to explain new phenomena, not only in mechanics but in electricity and magnetism as well. In the above quoted remark, Hertz, Helmholtz's student, explicitly admits that energetism was at bottom the more or less direct offspring of the success of the principle of energy conservation. Conceding that many other reasons also played in favour of *energetism* as a general representation of physics, Hertz had to motivate his different position on fundamental matters.

In his presentation, he chose Hamilton's principle of least action as the most suitable principle, among the integral principles of mechanics, to represent mathematically the energetist conception.[11] One of the advantages of this choice is that on "can deduce a whole series of relations, especially of mutual relations between every kind of possible force, which are wanting in the principles of the first image [i.e., that of customary mechanics]". These relations, he adds, were demonstrated by Helmholtz in his *Über die physicalische Bedeutung des Prinzips der kleinstens Wirkung*.

The energetist conception presents for Hertz (and for Kirchhoff) the other advantage of avoiding atomism as an ontology of physical reality:

> It best avoids talking about things of which we know very little......to an investigator like Gustav Kirchhoff who was accustomed to rigid reasoning, it almost gave pain to see atoms and their vibrations wilfully stuck in the middle of a theoretical deduction .[12]

Still other advantages are afforded by its simplicity and appropriateness:[13]

> In the hypothesis of the problem there only enter characteristics which are only directly accessible to experience, parameters, or arbitrary co-ordinates of the bodies under consideration....These are advantages in respect to simplicity, and hence of appropriateness.[14]

However, *energetism* also had defects which, in the end, led Hertz to reject it, among them, lack of CR and PM. In fact, Hamilton's principle cannot be applied to certain mechanical systems as, for instance, "when bodies of three dimensions roll on one another without slipping". A second defect is that it calls for an infinite value of potential energy, when one assumes an infinite total mass for the universe.[15]

These are defects of the theory and, as such, they might be reformable. In the use of "extremum" principles, like Hamilton's, there is, however, another difficulty: their implicit finality. This difficulty is of a philosophical nature and, as such, Hertz believes it to be of major significance:

> Hamilton's principle, has no simple physical meaning... it makes the present motion dependent upon consequences which can only exhibit themselves in

constructs and, as such, we are free in assuming different hidden quantities.[24]

In any case, admitting the existence, among the visible quantities, of hidden ones (entities), has the effect of changing the theory's status, from a description of nature to a mental representation (*Bild*), a model of phenomena.

Thus Hertz's epistemology in the *Prinzipien* reflects the insights he reached in his experiments on electromagnetic waves. In his 1891 introduction to the collection of his essays on his famous experiments, he presented different "modes of representation of Maxwell's theory" and he concluded that "the undying part of Maxwell's work" resides in the "common significance of the different modes of representation", which is symbolised in Maxwell's equations.

8.4 A Comparison between Helmholtz' and Hertz's Philosophies

Taking my thesis about the organic unity of Hertz's thought to a more general level, I would argue that his detailed discussion, in his introduction to *Electric Waves* of the reasons for his preference for Maxwell's theory over that of Helmholtz has to be read in parallel with his discussion in his introduction to *Prinzipien* of his preference for a space-time-mass (plus hidden quantities) conception of mechanics over energetism. In the first introduction, Hertz presented different "modes of representation" of Maxwell's theory and he concluded that "the undying part of Maxwell's work [resided in] the common significance of the different modes of representation" which is symbolised by Maxwell's equations. He also commented that

[In order to properly understand] how it is possible that ideas and conceptions which are akin and yet different may be symbolised in the same way in the different modes of representation... [and in order to have] a proper comprehension of anyone of these [representations] the first essential is that we should endeavour to understand each representation by itself without introducing into it ideas which belong to another.[25]

This exigency of order and clarity can only be satisfied by axiomatic analysis, precisely the method that Hertz took up again in his *Prinzipien*. There he defined the principles of mechanics as

[Any selection of proposition] such that the whole of mechanics can be developed from it by purely deductive reasoning without any further appeal to experience.[26]

He adds that

By varying the choice of the propositions which we take as fundamental, we can give various representations of the principles of mechanics.[27]

Hertz's fundamental equations of mechanics, being common to the various modes of representation, symbolise at its best the axiomatic structure of his mechanics. Let us compare Hertz's above mentioned ideas with his famous interpretation of the meaning of Maxwell's system of equations:

Maxwell's theory is Maxwell's system of equations. Every theory which leads to the same system of equations, and therefore comprises the same possible phenomena, I would consider to be a form or special case of Maxwell's theory....Hence in this sense and in this sense only, the two theoretical dissertations in the present volume can be regarded as representations of Maxwell's theory. In no sense can they claim to be a precise rendering of Maxwell's ideas. On the contrary, it is doubtful whether Maxwell, were he alive, would acknowledge them as representing his own views in all respects.[28]

I find a remarkable similarity between these two positions of

Hertz's substitutes for forces are *hidden quantities*, under the form of *hidden masses*, which are intermingled with other *visible quantities*, both in mechanics and in electrodynamics. They co-operate with th visible quantities in the description of motions by transforming all motions into inertial motions. Following the inclusion of hidden masses, potential energy, which is meaningless when forces are missing, can be redefined as simply kinetic energy of hidden masses themselves.[19] Energy conservation is thus reduced to Huygens's theorem, i.e., kinetic-energy conservation in an isolated system.[20] Let us note that Hertz's rejection of potential energy in 1893 echoes his 1890 rejection of vector potentials in his elaboration of Maxwell's theory.[21]

The emphasis placed by Hertz on the role of hidden masses, "confederates concealed beyond the limits of our senses",[22] indicate the importance he attaches to this innovation. The passage is worth quoting in its entirety:

> If we try to understand the motion of bodies around us and to refer them to simple and clear rules, paying attention only to what can be *directly observed*, our attempt will in general fail. We soon became aware that *the totality of things visible and tangible* do not form an universe conformable to law, in which the same results always follow from the same conditions. We become convinced that...*we are free* to assume that this hidden something is nought else than motion and mass again, motion and mass which *differ from the visible ones not in themselves but in relation to us and to our usual means of perception.*[23] [italics aremine]

The epistemological status of hidden masses is also a key for assessing Hertz's relationship to Kant. As regards their status, Hertz wavers between two conceptions: a) these quantities are hidden from our usual means of perception, not being visible and tangible; in other words, they are not directly observable, but, in themselves, they have the same status as the visible quantities; b) they are thought-the future, thereby attributing intentions to inanimate nature, but, what is much worse, it attributes to nature intentions which are void of meaning.[16]

It is surprising to discover that the vexed controversy on the alleged finality of the extremum principles, which today is usually considered to have been settled once for all in the eighteenth century, is still alive in Hertz's mind, since, for him, it was falsely settled by those who claimed it had a metaphysical character:

> The usual answer which physics nowadays keeps ready for such attacks is that these considerations are based upon metaphysical assumptions; that physics has renounced these, and no longer recognises it as its duty to meet the demands of metaphysics..... A doubt which makes an impression on our mind cannot be removed by calling it metaphysical; every thoughtful mind as such has needs which scientific men are accustomed to denote as metaphysical.[17]

8.3. *A Synthetic Representation of Mechanics*

Following the above premises, Hertz presented his "third arrangement of the principles of mechanics" as the one which he prefers above the others. Forces are not included among the fundamental conceptions, an assumption which is consistent with his earlier electrodynamical research. In that case he was forced by his logic of research to cancel electrostatic distant forces in favour of ether-polarisation and he saw no reason to keep the concept in his generalised mechanics. Besides, distant forces were always crucial for any theory aspiring to a lucid (*scheinbar*) model. The exclusion of force allows a significant gain in simplicity, a feature which Hertz rated highly:

> [This mode of representation] only starts with *three independent fundamental conceptions, namely those of time, space and mass* [my italics].[3]

Hertz. In fact, Hertz reaffirms his binary conception of the two aspects of theory: on the one hand, the different modes of representations and, on the other, their common significance, symbolised in the fundamental equations;

> The common significance of the different modes of representation (and others can certainly be found) appears to me to be the undying part of Maxwell's work.[29]

It is not far fetched to conclude that, according to Hertz, his fundamental equations of mechanics, sharing with electromagnetic equations the property of being common to the various modes of representation, symbolise, at its best, the axiomatic structure of physics.

I think that Hertz's rejection of the unitary role of *energetism* has to be justified by more fundamental reasons than the defects listed in the above passage. In effect, the third difficulty does not concern the energy theorem in itself but the selection of integral principles among other possible mechanical principles, and it could be overcome by other appropriate selections. Consequently, I find that a fundamental justification for Hertz's choice of his third arrangement is to be sought in the fact that this arrangement was consistent with his model-like *Bild* -conception of theory, because it offered a conceptual model that was absent in Energetism. Helmholtz claimed (see below) that the presence in Hertz's conception of this model, a "mechanical representation", was the main point of difference between his and Hertz's conceptions of theory.

The contrasting elements in Helmholtz's and Hertz's electrodynamic theories discussed above are also present in their epistemologies. Helmholtz believed that "the impact of a new abstraction can only be understood clearly when its application to the chief group of individual cases which it includes has been thought out and found valid".

His application of displacement to the Poisson-type polarisation has precisely the above characteristics, i.e., it is clearly understood in its application to an individual case. He added: "It is very hard to define new abstractions in universal propositions, so as to avoid misunderstandings of all kinds".[30] This type of abstraction, banned by Helmholtz, was exactly the one which induced Hertz to interpret Maxwell's displacement as hidden ether polarisation. It is worth noticing that, in Hertz's system, simplicity is reached at the cost of introducing concepts such as ether polarisation and concealed masses which have no correspondents on the empirical level.

It is interesting to read how the above contrast is seen by Helmholtz himself. In his preface to Hertz's *Principien,* Helmholtz remarked that his favourite student did not follow his master's philosophy when he privileged "mechanical representations" instead of the "simple representation of physical facts by systems of differential equations". In the same passage Helmholtz ranks Hertz with Kelvin and Maxwell among those physicists who have adhered to mechanical representations instead of following his representation of facts and laws as a simple system of differential equations:

> (The three physicists) have evidently derived a fuller satisfaction from...mechanical representation of electromagnetic processes...than from the simple representation of physical facts and laws in the most general form, as given in systems of differential equations.... For my own part I have adhered to the latter mode of representation (i.e., system of differential equations) and I have felt safer in so doing; yet I have no essential objections to raise against a method which has been adopted by three physicists of such eminence.[31]

The mechanical representations Helmholtz credits to Hertz as the characterising feature of Hertz's philosophy are to be related to

Hertz's *Bild* conception of theory as a theoretical model where concepts do not necessarily correspond to anything observable at the empirical level.³² In his dismissal of a correspondence between a concept and an observable, Hertz is consistent with his request that a theory represents a *Bild* or a *ScheinBild*.

CHAPTER 9

ON BOLTZMANN'S MECHANICS AND HIS BILD-CONCEPTION OF PHYSICAL THEORY

9.1. Boltzmann's Bild-*Conception an the Plurality of Theories in the 1890's.*

When, in 1892, Boltzmann published "On the Methods of theoretical physics",[1] he had read[2] Hertz's 1890 lecture on the relation between light and electricity, Maxwell's important works, "A Dynamical Theory of the Electromagnetic Field", and *A Treatise on Electricity and Magnetism*, as well as Maxwell's booklet, *Matter and Motion*. Concerning models, Boltzmann's main interest was Maxwell's method of "mechanical analogies", a method that Boltzmann accepted because in physical theory "the new approach compensates the abandonment of complete congruence with nature by the correspondingly more striking appearance of the points of similarity". Nevertheless he rejects any generalisation of Maxwell's ideas "that knowledge itself is nothing else than the finding of analogies" [3] and he also refuses to abandon completely the old method as supposedly "worn out in spite of all it has done".

Boltzmann's ideas on the scientific method are more thoroughly expressed in his 1899 Essay "On the Development of the Methods of Theoretical Physics in Recent Times".[4] Boltzmann had by now read[5] Hertz's *Prinzipien* (1894). His writings in the seven intervening years (1892-1899), published in *Populäre Schriften*, show his increasing interest in Darwin's ideas, his acquaintance with

Mach's philosophy and an attempt to build up more consistently his epistemological convictions.[6]

The 1899 essay was written as an address to a meeting of natural scientists at Munich on September 22, 1899, and it is conceivable that Boltzmann, on this occasion, wished to present his methodological ideas to this distinguished audience in the most general and comprehensive form. In this essay he shows a conciliatory attitude towards his philosophical opponents of the time, energetists and phenomenologists, by showing how their positions, if taken "cum grano salis" and without indulging in exclusiveness, could be reconciled with Maxwell's and Hertz's ideas and, implicitly, with his own.

Developing consistently with his aforementioned appreciation of Maxwell's analogical thinking, Boltzmann is now very sensitive to what he considers Maxwell's most important legacy, the plurality of theories:

> Maxwell warned against regarding a particular view of nature as the only correct one merely because a series of consequences flowing from it has been confirmed by experience. He gives many examples of how a group of phenomena can be explained in two totally different ways, both modes of explanation representing the facts equally well. Only on adding new and hitherto unknown phenomena does the advantage of one method over another reveal itself, though the former may have to give way yet to a third after further facts have been discovered [my italics].[7]

Contrary to the traditional conception of theory as a true description of nature (Boltzmann's "complete congruence with nature") or as a best approximation of it, theory is now presented by Boltzmann "as a mere representation (*Bild*)[8] of nature, a mechanical analogy as he [Maxwell] puts it, which at the present time allows one to give the most uniform and comprehensive account of the totality of phenomena".[9] Although this conception represents an undeniable failure with respect to the old descriptive conception of theories, at the same time, as Boltzmann himself comments, it has some advantages: the proliferation of theories, one of its consequences, is fruitful in "adding new and hitherto unknown phenomena". The abandonment of the descriptive conception is the price physicists must pay for this increased grasp of experiments, their heuristic fecundity, as we say today. The predictive power of a theory is not only enhanced, but it becomes one of theory's characteristic features.

Boltzmann's second important conclusion in 1899 is that the conception of the plurality of theories has among its consequences the rejection of the old criterion for theory-testing: the "crucial experiment". Reciprocally, the experimental confirmation of a theory cannot be considered a test of its "absolute correctness". Boltzmann gives, as an example of the latter, the Hall effect, predicted not only by Weber's "false" theory (i.e. magnetic action is effective not just on the current-carrying wire but also on the particles which constitute an electric current), but also by Maxwell's "true" theory [10]. This second 1899 conclusion will be rejected by Boltzmann in his final Darwinian epistemology, as I show below.

In 1899 Boltzmann considered Hertz's philosophy an advance in the direction opened up by Kirchhoff and Maxwell. He believed that Hertz had deepened philosophically Maxwell's epistemological ideas:

In his book on mechanics Hertz has given a certain completion not only to Kirchhoff's mathematico-physical ideas but also to Maxwell's epistemological ones.... Hertz makes physicists properly aware of something philosophers had no doubt long since stated, namely that no theory can be objective, actually coinciding with nature, but rather that each theory is only a mental picture [eingeistiges Bild] of phenomena, related to them as sign to the designatum [my italics].[11]

With this view of theory Boltzmann then concludes:

> The question whether matter consists of atoms or is continuous, reduces to the much clearer one, whether the continuous is able to furnish a better picture of phenomena.[12]

This remarkable statement concisely expresses the essence of Boltzmann's 1899 view on atomism: it is not necessary to accept atomism in an ontological sense, it suffices to use atoms with the aim of achieving new theoretical views, as Clausius did in his kinetic theory.[13] Being no longer an "ontology", atomism is not a necessary hypothesis in physics, nor the only one which can be fruitful. In fact, according to Boltzmann, Ampére, Franz Neumann and Kirchhoff did not base their derivations on molecular ideas, even if they did not deny the atomistic structure of matter .[14]

Boltzmann's refutation of atomism as an ontology is consistent with those ideas which for him represented (in 1899) the link between Hertz and the phenomenologists. More precisely: atomism can be reconciled with phenomenology on the one hand and, on the other, with Hertz's and Maxwell's ideas provided it is taken as a construct, a *Bild*, (a mental model) in Hertz's sense. Atomism is an important example of this very feature of the *Bild*, precisely that of going beyond experience. Atomism in fact: 1) is consistent with the idea of Theory as a "mental model" (*Bild*), 2) is an indispensable condition for prediction.

These two latter features of a *Bild* are correlated in Boltzmann's mind :

> [It] follows, from the nature of the intellectual process itself, consisting as it does in adding something to experience and creating a mental picture [geistiges Bild]that is not experience and therefore can represent many experiences. Only one half of our experience is ever experience, as Goethe says. The more boldly one

goes beyond experience [*über die Erfaharung hinausgehet*], the more general the overview one can win, the more surprising the facts one can discover, but the more easily too one can fall into error. Phenomenology therefore ought not to boast that it does not go beyond experience, but merely warns against doing so to excess.[15]

In the last passage Boltzmann succeeds in obtaining a remarkable balance between the opposite pretensions of crude phenomenology and crude (ontological) atomism. Notice that this balance is a direct consequence of Hertz's and Boltzmann's conception of a theory as a mental representation [*Bild*]. In fact this conception contradicts the naive atomist's pretension that atoms do really exist in nature and, at the same time, it contradicts the phenomenologist's pretension that a theory should just describe phenomena. Reciprocally, analogical-atomism can be accepted by phenomenologists because of its predictive power and that degree of phenomenology which is implicit in any analogical model can be tolerated by atomists as a warning against any going beyond experience to excess.

I intend now to illustrate the following points: 1) Boltzmann's latter conciliation of Mach and Hertz is reached at the cost of modifying some of the features of Hertz's *Bild* conception. 2) The aforementioned balance is somewhat revised by Boltzmann in his final writings, in favour of a tempered prudence. Specifically, his above encouragement to go beyond experience ("über die Erfaharung hinausgehen") will later be tempered by his fear of "a genuine mistake" in the "deductive method".

9.2 Boltzmann's Mechanics and his Opposition to Hertz's *Zulässigkeit*

Boltzmann did not learn from Hertz his philosophy of *Bilder* but he was exposed to an intense training in the post-Kant philosophy in his university years.[16] However, in his *Populäre Schriften* he almost always contrasted his own view on *Bilder* with Hertz's ideas. There is then no better way of grasping Boltzmann's opposition to the a-priori requirements on Hertz's *Bild* then to follow Boltzmann's line of thought both in his adherence to and his criticism of Hertz's ideas. In "On the Fundamental Principles and Equations of Mechanics", which presents the content of the first two[17] of four lectures given at Clark University in 1899, Boltzmann writes:

> All our ideas and concepts are only internal mental pictures [*innere Gedankenbilder*], or if spoken, combinations of sounds.... We can therefore pose such formal questions as whether only matter exists and force is a property of it, or whether force exists independently of matter... *but none of these questions are significant* since all these concepts are only mental pictures whose purpose is to define [*darstellen*] phenomena correctly. This was stated with special clarity by Hertz in his famous book on the principle of mechanics, except that he there begins with the demand that the pictures we construct for ourselves must obey the laws of thought. *Against this I should like to urge certain reservations* or at least to explain the demand a little further. Certainly we must contribute an ample store of laws of thought, without them experience would be quite useless, since we could not fix it by means of internal pictures. These laws of thought arealmost without exception innate [*fast ausnahmlos angeboren*], but nevertheless they suffer modification through upbringing, education and our own experience [my italics].[18]

Boltzmann refers here to the central aspect of the problem: the meaning of the expression "law of thought" (*Denkgesetz*). In his effort to introduce a distinction between his and Hertz's *Denkgesetzen*, Boltzmann gives examples where these laws have been falsified by empirical knowledge: one example is the geocentric theory, with its absolute conception of the antipodes, etc. Such conceptions "at the time were regarded as self-evident laws of thought, whereas we are now convinced that they are futile".[19] It is evident that for Boltzmann, laws of thought are identified with empirically testable laws, parts or components of *Bild* Theory:

I therefore wish to modify Hertz's demand and say that insofar as we possess laws of thought that we have recognised as indubitably correct through constant confirmation by experience, we can start by testing the correctness of our pictures [Bildern] against these laws.[20]

Laws of thought and consequently *Bilder* seem to diminish in their epistemological status as Boltzmann's ideas evolve to their ultimate form. In his Address[21] to the St. Louis Congress of 1904, he expresses the opinion that casual connection is not at all a regulative principle:

Indeed people racked their brains over the question whether cause and effect represent a necessary link or merely an adventitious sequence, whereas one can sensibly ask only whether a specific phenomenon is always linked with a definite group of others, being their necessary consequence, or whether this group may at times be absent.[22]

According to Boltzmann, a causal link is just "a mode of action" that is "appropriate in most cases", but "if somewhere it ceases to be appropriate" it cannot be relinquished because it has become "so habitual" and a "second nature". In such cases where "laws of thought" contradict "the world" there is evidence that "adaptation overshoots the mark (Anpassung schließt über das Ziel hinaus)". An evident attack on Kant's doctrine closes the passage:

We must not aspire to derive nature from our concepts, but must adapt the latter to the former. We must not think that everything can be arranged according to our categories [*Kategorien*] or that there is such a thing as a most perfect arrangement.[23]

As to the question of whether there is or is not a limit to the divisibility of matter, a theme that Kant "discussed in his antinomies... and explains that both the case for and against can be proved by strict logic", the question is, according to Boltzmann,[24] "devoid of sense and hope" when asked in general. It has, however, meaning in a special case, the one related to the cogent problem of atomism:

Which represents the observed properties of matter most accurately, the properties on the assumption of an extremely large finite number of particles, or the limit of the properties if the number grows infinitely large?[25]

In answering this question he writes:

We shall... on the one hand, start only *from what is given,* while, on the other, in forming and linking our concepts we shall heed only the aim of obtaining as adequate an expression as we can of *what is given.* [my italics].

"What is given", the empirical strength of data, seems in 1904 to be Boltzmann's major if not the only concern in deciding between theories, a statement which contradicts somewhat his 1899 rejection of experiments as having a "crucial" bearing on theory. Overturning Hertz's methodological advice, Hertz's "correctness" is to be preferred to Hertz's "permissibility", the non-empirical criteria of an "inner perfection". Boltzmann, by his own admission, renounced this inner perfection in his mechanics:

The picture [of mechanics] that Hertz thus constructs independently of experience has a certain *inner perfection* and obviousness and contains really only very few arbitrary elements. As against this my picture is evidently inferior, for it contains many more features that are marked by an *absence of inner necessity,* being introduced only in order to facilitate subsequent agreement with experience [my italics].[26]

Consistent with his epistemological tenets, Boltzmann's mechanics has an axiomatic foundation more akin to the traditional formulation of the Newtonian-Laplacian theory. As a consequence it misses contiguous action, which, according to him "however a-priori it may seem to some, still *goes completely beyond the facts* and to date remains well beyond what can be elaborated in detail" [my italics].[27]

To the lay reader it might seem strange that, more than ten years after Hertz's celebrated experiment on the propagation of electromagnetic waves, Boltzmann still considers contiguous action as "completely beyond the facts". Yet Hertz himself would have agreed that no crude fact could directly show contiguous action, but he would however have contended that only an a-priori (PM observant) theory could give support for contiguous action to be shown by an experiment (see above chapter seven).

While Hertz believed (see above chapter seven) that his mechanics with its hidden masses had opened a path for a solution to the problem of the mechanical explanation of electrodynamics,[28] Boltzmann acknowledged the contemporary failure of mechanical representation of electrodynamics, although he still hoped that such a representation will be achieved in the future:

At most we can say that for certain phenomena an attempt at mechanical explanation is as yet premature. The general question as such can be resolved after centuries, or at least be given a new setting and clarified.[29]

Mechanics is here defined as "the theory of the simplest phenomena without which no others could be conceivable", and as "the basis of natural science as a whole, all other scientific theories resting on mechanics". Determinism is, at bottom, the basic concept of Boltzmann's conception on mechanics:"a precondition of every scientific insight is the principle that natural processes are unambiguously determined; or, in the case of mechanics, that motions are".[30] Boltzmann believes that his own century is "the century of the mechanical view of nature, the century of Darwin". Apriorism is equated by him with biologic innatism:

One can call these laws of throught a-priori because through many thousands of years of our species' experience they have become innate to the individual, but it seems to be no more than a logical howler of Kant's to infer their infallibility in all cases. According to Darwin's theory this howler is perfectly explicable. Only what is certain has become hereditary; what was incorrect has been dropped...[31]

Notice the sense in which Boltzmann uses terms such as "mechanics", "mechanical":[32] "the most splendid mechanical theory in the field of biology, namely the doctrine of Darwin".

There is good evidence that Boltzmann's criticism of Kant's philosophy increases in strength in proportion to his enthusiasm for philosophical evolutionism. Similary, Boltzmann's enlarged conception of the meaning of the science of mechanics grows apace:

In nature and the arts the all-powerful science of mechanics is thus ruler, and likewise in politics and social life.[33]

Coming to the problem of the role of axiomatics, Boltzmann also considers logical clarity to be the only valuable criterion in theory construction, not to be compromised, for fear that our constructs will be arbitrary, i.e., empirically empty, because of a rush to bring in "experience too early".[34] However, as I have shown, in 1905 he had reached the conclusion that in theory construction PM has not to be preferred "per se". In other words, he does not consider axiomatic structure to possess any intrinsic value (a regulative role) in Hertz's sense. Hertz's evaluation of PM consistently leads to the conclusion that any theory can be tested only as a relational whole (see above chapter six). The only judge of theory is, ultimately, experience; both Hertz and Boltzmann would agree on this statement. However, unlike Hertz, Boltzmann concludes that separate laws can be tested as such (i.e., piece by piece) against experience.[35] Boltzmann arrived at this conclusion in his later years, overcoming those doubts on the crucial nature of experiments that, as I have shown, he expressed in his 1899 essay.

9.3 The Historical Role of Boltzmann's Mechanics and Gas theory

In trying to reconstruct the main line of development in Boltzmann's philosophical ideas concerning PM, my ultimate concern was to investigate if and how these ideas were related to his theoretical contribution to mechanics and gas theory. As regards mechanics, he himself has given evidence, as we have seen, that the special structure of his mechanics was directly related to his rejection of Hertz's PM in its primary role: this is one case, at least, where Boltzmann's philosophy coincides with Boltzmann's physics. His protracted defence of his kinetics theory against the attacks at Loschmidt and Zermelo is, as is well known, an important aspect of his strategy for supporting his faith in mechanics. Here also, his philosophy meets his physics, at least indirectly, i.e., through his mechanistic faith.

I wish to attempt here to find some more direct traces of how Boltzmann's philosophy influenced his gas theory. I refer to Boltzmann's doubts about his faith in mechanics and his prevarication between pure mechanical and probabilistic conceptions, noted by many of Boltzmann's historians. Perhaps the clearest example of Boltzmann's wavering was given by Thomas Kuhn. According to

Kuhn, Boltzmann, in his much discussed 1872 proof of the H theorem, thought that "his results were of the deterministic sort", i.e., " an apparent proof from mechanics that a gas must evolve to equilibrium from an arbitrary chosen initial state".[36]

However the probabilistic or combinatorial treatment of irreversibility (*Komplexions Rechnung*) developed by Boltzmann in 1877 is very different from his 1872 approach. A third "conceptual cluster" is briefly presented by Boltzmann, according to Kuhn, in thegfirst volume of his 1894 work on gas theory, centred on the notion of molecular disorder and consisting in the hypothesis of "a prohibition of certain actual configurations of the molecules within individual cells, configurations which the laws of mechanics, taken alone, would otherwise allow".[37]

The place where "for the first time Boltzmann had tied an apparently fully probabilistic statement directly to a discussion not ot the combinatorial but of mechanical H-Theorem, was his letter to *Nature* of February 1895".[38] In this letter, Boltzmann attributes a probability measure to a value of his H function. In sum, according to Kuhn, Boltzmann never completely abandoned in the course of his scientific career, a deterministic (not probabilistic) conception of gas theory, although probabilistic conceptions were popping up, here and there in his scientific works.

In his attempts to overcome generalised mechanics a central role is played by his complexion law, the famous $s = k \, log \, w$, (it appears on his Vienna Memorial) which expressed mathematically his new concept of the probability of a state. In this theory, the traditional (event-centred) concept of the probability underwent a radical upheaval, and a new concept appeared on the physics scene. As is known, Boltzmann's complexion law guided Planck towards one of the most radical innovations in physics, quantum of action in Planck's black-body radiation law.

In his now classic work, Max Jammer points to the fortunate circumstance whereby Planck repeatedly turned to a Boltzmann-type formula for entropy, neglecting other more traditional approaches such as those via the equipartition theorem. At the same time Jammer stresses the fact that Planck trusted the pure formal content of Boltzmann's law, ignoring its physical content perhaps because "of his profound aversion to the molecular approach". Jammer emphasises still that, whatever the motivations that led Planck to write an unexpected relation for his entropy:

Planck ultimately found it necessary to abandon his «thermodynamic approach» and to turn to Boltzmann's probabilistic conception of entropy... *apparently following Boltzmann...* [when writing]... SN=k log w... [my italics].[39]

However different Planck's procedure was from Boltzmann's in computing the number of complexions in his Boltzmann-type formula for entropy, one thing is certain: the formal structure of Boltzmann's law and not its physical content guided Planck towards his achievement. The same formal structure of this equation channelled, in its turn, classical statistics towards the new-statistics. The latter derived its origins, as is known, from the Bose-Einstein contributions and emerged by a non-linear path through the works of Planck, Debye, Ehrenfest, De Broglie, Einstein and others.

Silvio Bergia has shown[40] how the fundamental innovation in the new statistics, atomic-indistinguishability, was implicitly contained (although Bose was unaware of it) in Bose's handling of Boltzmann's celebrated formula. Again one has to underline that it was the formal structure of this formula which, in a differentiated context of conceptions, forced almost *malgré soi* Planck and Einstein to accept the indistinguishability concept. If, as M. Delbruck puts it,[41] "serendipity" has to be invoked in order to explain the unexpected conceptual jump of indistinguishibility a jump made from many sides "serendipity" was, in this case, guided by the formal structure of Boltzmann's theory. Is this not a success for those formal requirements on theory stressed by Hertz's PM?

I find that Boltzmann's rejection of Hertz's PM has a deep justification in his mature epistemology: in the cultural struggle of his time between a neo-Kantian tradition and a self-affirming Darwinism, he stood for the latter. As shown above, determinism was at the base of his conception of generalised mechanics and he thought he could find in generalised mechanics the necessary conceptual weapons to counter the attacks of both atomists (such as Loschmidt and Zermelo) and anti-atomists, such as Mach and the energetists. Against Loschmidt and Zermelo who proved that his mechanical gas theory was leading to paradoxes, Boltzmann tried to vindicate his mechanical results at the cost of introducing mechanically unjustified probabilistic hypotheses. To Mach and the energetists, who claimed that atomism was hypothetical and unjustified in the theories of physics, he wanted to show how his approach to gas theory via atomism was experiencing success and likely to produce more. At this point any concession to the Hertzian neo-Kantian *Zulässigkeit* would have weakened Boltzmann's opposition against both types of opponents.

Boltzmann's profound intellectual insight and his open-mindedness, both a consequence of his philosophical and cultural interests and studies, suggested to him possible directions for overcoming his peculiar form of generalised mechanicism and Darwinism. In his article in *Nature* (1895), he explicitly states that "my Minimum Theorem, as well as the so called second law of Thermodynamics are only theorems of probability. The second law can never he proved mathematically by means of the equations of dynamics alone". (One should however keep in mind that probability calculus can be considered as a (Laplacian) method of circumventing mathematical difficulties in mechanical equations;[42] hence Boltzmann's statement does not definitely disprove his mechanical ideas). After the appearence of Lorentz's and Abraham's electron theories, he somewhat prophetically stated in 1904 that an atomic theory could dispense with his mechanistic tenets:

The ray of hope for a non-mechanical explanation of nature came not from energetics or phenomenology, but from an atomic theory that in its fantastic hypotheses surpasses the old atomic theory as much as their elementary structures surpass in smallness those of the old atoms. I need not mention that I mean the modern theory of electrons.[43]

Finally we should recognise that Boltzmann's strenuous struggle to underpin his conception of mechanics contributed fruitfully to the historical process of clarifying for us how mechanics and statistics overlap and differ. That he was somehow aware of participating in this process is hinted at in an otherwise obscure statement. It appears as one of his comments on Planck the editor referred to and his intellectual opponent[44] on the occasion of the publication of Kirchhoff's Lectures on the theory of heat:

Even those who like the editor of the [volume] now under discussion think gas theory unworthy of the acumen expended on it, would not wish those who do write on the subject to expend less.[45]

According to my reading of Boltzmann's statement: deterministic gas theory is worthy of great scientific effort notwithstanding its being an unworthy subject. Perhaps the conception of science as a process, in its historical dimension, cannot be better confirmed than by this statement. Elsewhere[46] Boltzmann considered himself also "as an unskilled labourer whose task was to level off the way to the building, to clean the facade and perhaps to insert here and there some foundation stones". The building is not yet completed and we, part of the process to which Boltzmann so skilfully contributed, still insert here and there those foundation stones.

PART THREE

From Relativity to Quantum Theory

FOREWORD TO PART THREE

Part three deals with the epistemologies which underly Einstein's and Bohr's controversial positions on the foundational assumptions of their theories.

Although often neglected by historians, the nineteenth-century tradition of theoretical physics and the problems on the theory-experiment relation, discussed in the pages above, are important and must be taken into account in the historical reconstruction of Einstein's thought. The epistemological discussions of Helmholtz, Boltzmann, and Hertz are part of Einstein's cultural heritage. As is well known, Einstein was acquainted with the works all the physicists examined above. He often expressed his own understanding of the theory-experiment relationship.[1]

I argue that, following the success of relativity and quantum Mechanics, physicists were confronted with a very great challenge to the continuity of their scientific enterprise. Continuity is a basic requirement of an empirical science: our perceptions have a historical dimension because they include the memory of past events (in particular, theories and experiments). Our perceptive space, which is usually deemed as synchronous, has also a temporal extension and a temporal continuity.

This fact might explain why scientists, such as Thomson and Maxwell, much of whose research was highly innovative, always considered their innovations as improvements on previous theories.

For this reason, Maxwell considered theoretical pluralism, which might be taken as contradicting temporal continuity, as a provisional position (see above chapter four).

Continuity in the development of theories implies that a new theory should reproduce, under certain conditions (the so-called limiting case), the sound parts of the theory it intended to supersede. Einstein attributed great importance to this feature of the new theories, and Bohr explicitly elevated it to the role of a principle. According to Max Jammer,[2] the idea that quantum theory (QT), or at least its formalism, contains classical theory (CT) as a limiting caseis first stated by Planck in 1906, in his well-known statement that when h tends to zero his radiation formula goes over into the classic Rayleigh-Jeans formula. A correspondence criterion was implicitly contained in this statement but, to my knowledge, its conceptual premises were not explicitly formulated prior to Einstein and Bohr.

It is therefore meaningful to say that Einstein explicitly announced the continuity requirement as a correspondence criterion.

Although, by common agreement, Einstein's general theory of relativity (GR) was considered one of the most revolutionary theories in the history of physics, his desire to be faithful to the doctrine of continuity may be understood to indicate his awareness that a revolutionary theory might break too radically with previous theories.

In my essay on Einstein's notion of correspondence, I explore Einstein's technical usage of a correspondence criterion (CCr) as a guide towards generalising special relativity. I also refer to his thoughts and perplexities in the face of the above mentioned doubts posed by the significance of correspondence.

The continuity problem also intrigued Bohr at a certain

period in his research, because classical theories (CT) represented a glorious patrimony of science which had to be recovered and salvaged in some way.

In my chapter on Bohr's correspondence and complementarity principles , I look at the historical development of the two principles in Bohr's various contributions to the Copenhagen philosophy.

The significance of Bohr's complementarity for quantum theory has been studied by historians and epistemologists of high renown and has prompted the comments of a philosopher of the calibre of Karl Popper.

My point is that complementarity can be considered a coherent continuation of Bohr's ideas on correspondence, and I share with some historians the conviction that the positions of Bohr and Heisenberg on the epistemology of quantum theory are to be considered as a way of circumventing rather than solving the difficulties in relativity theory which worried Einstein.[3] Similarly, Einstein did not solve Lorentz's difficulties with the ether-drag experiment but, as shown in Holton's classical study, completely changed the way of approaching the problem.

I argue that Einstein's and Bohr's conceptions of the nature and role of theory should be viewed in the context of their responses to the problems posed by the development of theoretical physics in our century.

Another response was given by Schrödinger. The lack of individuality of the atomic particles presented in the new statistics, and in Heisenberg's indeterminacy relations, were considered by Schrödinger to be aspects of a more general crisis in the ontology of classical atomism.

Unlike his 1926 ideas, he now proposed to represent the wave equation in an n-dimensional space, and he considered second-quantisation technique to be the proper mathematical tool for his new conception of Physics.

Although Schrödinger accepted that space-time discontinuities and casual gaps may appear here and there on the observational level (e.g., in the indeterminacy relations), he was convinced that they could be made compatible with a continuous pure theory, provided one accepted a suitable conception of the theory's epistemological status. For him, only a continuous theory satisfied the conditions for a *complete theory*.

On these matters he thought he was adhering to the ideas of Hertz and Boltzmann, which were also reflected in the teaching of his master Exner. He sometimes referred to the "completion of experience in thought", a view that he attributed to Mach.

CHAPTER 10

EINSTEIN'S CORRESPONDENCE CRITERION AND THE CONSTRUCTION OF GENERAL RELATIVITY

10.1. Correspondence as a Warranty of Continuity with Tradition

The exigency of establishing a continuity with tradition by showing that, even when introducing radical innovations, new theories do not break with nor contradict well-confirmed former theorie is so widespread in modern science that it can be considered one of its main postulates. This postulate represents scientists' trust in the objectivity of the natural world and their presumption that, by affording a more or less accurate description of the world, physical theories cannot be contradictory.

In this sense the above exigency is a manifestation of the realist view of science. It is understandable that the more a new theory presents itself as highly innovative let us say revolutionary, in the sense of breaking with the past the more the foregoing realist view risks being twisted and endangered.

Since, by common agreement, Einstein's general theory of relativity (GR) was considered to be one of the most revolutionary theories in the history of physics, the story of his struggle to show that GR was somehow related through a correspondence criterion (CCr) to the spacial theory of relativity and Newtonian gravitation appears at times as a struggle with dramatic overtones. In any case, it should be conceded that the role of CCr within GR receives indirectly great conceptual significance.

The realistic views above are not, however, to be considered as necessary conditions for accepting CCr, for its fruitfulness may be justified from several perspectives. This point might explain why the realist views were not by any means shared by all physicists who fruitfully used CCr in the development of their theories: the above strongly realist positions were not shared by Einstein, nor did they become more acceptable to Bohr in his struggle with CCr (I refer to my discussion of Bohr below).

In this chapter I show that, in Einstein's case, CCr represented more than anything else the technical form of a general criterion for generalising the theory, in the rapid development of such a highly mathematicized discipline as GR. In this role CCr guided the theory's development in many remarkable ways, which I shall examine in the following.

10.2. What does Correspondence really mean?

In order to free our enquiry from preconceived ideas about the way theories of physics should develop, let us briefly examine some of the problems involved in defining CCR.

Correspondence (Cr) is usually defined as "the condition that the new, more comprehensive theory should relinquish the sound parts of the theory it intends to supersede".[1]

Although the definition above, and others similar to it, may be accepted as expressions of a suitable qualitative statement for an elementary treatise on physics, they prove to be of little help, if not completely circular and inconclusive, when critically analysed. The reason is that, in these definitions, the term "more comprehensive" to a theory is not clearly defined independently of the correspondence criterion itself.

In fact, a "more comprehensive theory" can be defined as that which includes a larger number of laws than a less comprehensive one. The definition above, founded as it is on a term-to-term comparison of laws, presupposes the possibility of *isolating* in a *given theory* either a) a single law or b) a set of such laws in order to compare them with the laws of the other theory and so to decide which of the two sets is the larger one. But this possibility is denied by the now commonly accepted Duhem-Quine thesis that theories consist of a logically connected group of propositions such that the isolation of either a) or b) changes their axiomatic structure and their semantics.

In order to further pursue our analysis let us consider a more detailed definition of Cr:

> The operational equations of a new theory must reduce, within the appropriate accuracy, to the corresponding operational equations of a well-established previous theory in the "regions" where the previous theory is well supported by data...The term "region" is to be interpreted broadly, not solely as a geometric region, but as a region of values of some *appropriate parameters* such as observed speed, size of objects, density of matter, temperature, or others. These parameters can be considered to be the independent parameters.[2]

Although Cr is here limited to operational equations, the same author of the passage above admits that it is very difficult, if not altogether impossible, to define operationally all the terms in an equation. He also adds that "the various terms in a scientific theory have explanatory value which go beyond their operational uses".[3] This is precisely the problem Einstein met in using Cr as a guide for GR construction, as we shall see in the following.

Other ambiguities in the above definition concern the selection of the *appropriate* parameters because the doubt remains that one selects such quantities on the basis of how far they allow Cr to hold.[4]

The considerations above do not forbid the mapping of a subset of an axiomatic system on a set of another such system, for everybody knows that the set of natural numbers can be mapped on to the larger set of rational numbers. But it is different to maintain that this operation implies that the correspondent entities are identical, just as it is clear that the natural 5 is not identical with the rational 5.0000000000....: However it seems to me that this type of identity was the one Einstein postulated when he required that his physical definition of the rod-length in special relativity could be transferred to GR (see the following).

Other difficulties with Cr emerge from the fact that theories such as GR and quantum mechanics (QM), when examined from a Cr perspective, present the interesting feature of reproducing not a single but a number of different, less comprehensive theories.[5] This occurs when different quantities can be selected as characteristic parameters and are given limit values. Einstein was also confronted with this case of a *pluricorrespondence* (PCr) when in GR he got as sub-theories special relativity, the Newtonian and the linearized gravitational theories.[6]

The difficulty is that what is really lost with PCr is the possibility of having a criterion for ordering the various theories in a series of growing comprehensibility (the criterion proper expressing for us the true meaning of CCr). Because, if the various correspondent sub-theories are not in themselves in correspondence, i.e., capable of being ordered according to the comprehensibility criterion, any ordering criterion is lost, and with it the meaning of Cr itself.

However, in spite of the above conceptual difficulties, the historian should acknowledge that CCr played the most important and fruitful role in the construction of GR, and, more generally, in the development of physics. This fact represents an interesting theme for future historical research.

10.3. The Heuristic Role of CCr in Einstein's 1912 Construction of GR and the Relinquishing of the Absolute Equivalence Principle

A consistent and almost systematic use of Cr in Einstein's construction of GR can be observed in his 1912-13 contributions. In 1913, in connection with his evaluation of Max Abraham's theory, he explicitly stated four postulates which any new relativistic gravitational theory should validate, although not necessarily all of them at one time. The third postulate concerned "the validity of the relativity theory (in the restricted sense); i.e., the system of equations are covariant with respect to linear orthogonal substitution (general Lorentz transformations)".[7] The third postulate was clearly a special application of CCr because special (or restricted) relativity corresponds to a situation of null gravitation.

He conceded that adherence to the postulate above was not compulsory. If, in return, other advantages were offered by the theory, this postulate, and Cr with it, could be abandoned. But, for him, this was not the case with Abraham's gravitational theory. He could not tolerate from this theory the omission of restricted Relativity as a special case. In conclusion, he formulated a general rule concerning adherence to postulate No.3 and CCr:

In my opinion, one has to keep faithful to Postulate 3 unconditionally as long as one does not find *cogent reasons* to abandon it. As soon as we abandon it we meet an infinite variety of alternatives [Italics added].[8]

In 1912 he was faithful[9] to this self-prescribed rule, in his attempt to lay down the foundations for a relative theory of a static gravitational field; he wrote the following equation for gravitational potential :

$$\Delta c = kc\rho \quad , \tag{1}$$

$c = c(\phi)$, the velocity of light, a function of the gravitational potential ϕ, K the gravitational constant, and ρ the density of matter.

The form of the equation is evidently modelled on the Newton-Poisson equation for gravitational potential ϕ :

$$\Delta \phi = K\rho \quad . \tag{2}$$

Among the reasons which could have convinced him to represent a gravitational potential by the velocity of light, one is certainly related[10] to the consequences on the variation of light velocity of his initial formulation of theequivalence principle (EP) in 1907.

He modified equation (1) in March of the same year when he found that it did not satisfy the laws of energy and momentum conservation. The new equation is:

$$\Delta c = k(c\sigma + \frac{1}{2k}\frac{(\nabla c)^2}{c}) \quad . \tag{3}$$

The new equation was no longer linear in the gravitational potential c. Moreover the density σ resulted from a superposition of the usual ponderable matter ρ and of the density of the electromagnetic field. The non-linear term in c was added to account for the contribution to the field sources themselves of field energy: a feedback effect in accordance with the mass-energy equivalence of the former special theory.

However, a new difficulty arose because of the non-linearity of equation (3): the new equation was inconsistent with both CCr and the 1907 equivalence principle (EP).[11] The first inconsistency can be proved by considering that the same equation, in the absence of ponderable matter and fields, takes on the form:

$$\Delta c = \frac{(\nabla c)^2}{2c} \ .$$

It corresponds to the Laplacian equation only in infinitesimal space-time regions, in which second space derivatives of c can be ignored. But this implied in turn that EP could remain valid only limited to such an infinitesimal domain (local EP).[12] (Although Einstein did not publish it, a demonstration along the lines above can be found in his Zurich notebook).[13]

In March 1912, Einstein declared that he adhered to the EP locality with some difficulty:

> My decision to make this step was not taken without difficulty, because thereafter I abandoned the ground of an absolute EP. It seems to me that an EP can be maintained only for an infinitesimal field.[14]

Einstein used local EP in 1913 and in the later development of the tensorial theory. However, although the necessity of transforming absolute EP into a local EP was amply proven in the new tensorial theory, he never renounced his belief in an absolute EP, as Norton has well illustrated.[15] Perhaps the latter was for him related to a powerful application of CCr in his initial generalisation of special into general relativity.

10.4. CCr Required a Sacrifice in 1913: Einstein's Provisional Relinquishing of General Covariance

Up to this point, Einstein's observance of his self-prescribed rule had resulted in some success. In 1913 he became convinced that a scalar theory could not be reconciled with general covariance and he and Grossmann shifted their program in the direction of a tensorial theory. As is known, they soon met an impasse in reconciling CCr with the general covariance of the new field equation.

Let us note that general covariance represented the core of Einstein's program in GR. It is the more surprising then that he abandoned general covariance to remain faithful to CCr. As his intellectual opponents, with Max Abraham in the forefront, immediately remarked, this abandonment amounted, at bottom, to a repudiation of Einstein's initial general covariance program in GR, one that his enemies likened, not without malice, to his abandonment at the start of his GR program of the Lorentz invariance of his former special relativity.

As has been illustrated,[16] Einstein's decision was the consequence of his and Grossmann's choice of the Ricci tensor as the mathematical expression of the gravitational field tensor. In fact, in order to satisfy CCr, the new equation should have, in analogy with the Newton-Poisson equation, a definite form of the type:

$$\Gamma_{\alpha\beta} = k\theta_{\alpha\beta} \quad , \quad \alpha, \beta = 1, 2, 3, 4 \quad , \quad (4)$$

$\Gamma_{\alpha\beta}$ is the Ricci second rank tensor of the gravitational field, built up through the symmetric metric tensor $g_{\alpha\beta}$. The energy-momentum tensor for matter is $\theta_{\alpha\beta}$.

Since the classical Newton-Poisson equation contains second order derivatives of the field, in order to establish a CCr between the classical potential ϕ and the above metric tensor $g_{\alpha\beta}$, it was necessary that the latter contained at most second order derivatives of the said metric tensor.

In the Cr area, it was assumed that fields were weak and static and the limit process was then performed. Firstly it was stated that Cr should hold for weak fields, when second degree terms could be neglected, i.e., when $g_{\mu\nu}, {_\alpha}g_{\tau\lambda, \beta} \to 0$. In this case we have a Ricci tensor of the form:

$$\Gamma_{\alpha\beta} \cong \frac{1}{2} g^{\tau\lambda} (- g_{\alpha\tau, \lambda\beta} + g_{\tau\lambda, \alpha\beta} + g_{\alpha\beta, \tau\lambda} - g_{\beta\lambda, \alpha\tau}) . \quad (5)$$

Then, the static field limit is applied. At that time Einstein was convinced that the *correspondent field* should have assumed the form it had taken in his former 1912 static field theory, which amounted to the most simple generalisation of special relativity. According to this special choice of the correspondent field,

$$g_{\mu\nu} = diag(-1, -1, -1, c^2(x^i)) \quad i=1,2,3 , \quad (6)$$

i.e., $g_{44, i} \neq 0$ is the only variable component.

With this CCr, the equation (5) becomes:

$$\Gamma_{44} = \frac{1}{2} \Delta c^2 \quad , \quad i,j=1,2,3 ,$$

$$\Gamma_{ij} = - \frac{1}{2c^2} \frac{\partial^2 c^2}{\partial x_i \partial x_j} , \quad (7)$$

$$\Gamma_{4j} = 0 \quad .$$

Due to the requirement that, without sources, $\Gamma_{ii} = 0$, the only admitted solution is when c^2 is a linear function[17] of the space co-ordinates x^i. But the Ricci tensor loses its generally covariant properties under these restrictions.

The impossibility of retaining the validity of both a generalization of a Ricci tensor (equation (5) above, implying CCr) and its general covariance, was expressed thus:

> The result...is that it is impossible to find for $\Gamma_{\alpha\beta}$ a differential expression which is both a generalisation of $\Delta\phi$ and also a tensor-like expression under a transformation whatever[18]

Faced with the alternative of rejecting either CCr or general covariance, Einstein rejected general covariance, keeping faithful to CCr.

I find that on this occasion Einstein transgressed his self-imposed rule above. In fact, faithfulness to this rule would have implied retaining covariance because covariance (i.e., a generally invariant tensor-like expression for $\Gamma_{\alpha\beta}$ above) represented a really cogent reason. His rejection of covariance in favour of CCr was only temporary. In fact, later on in 1916, he reinstated covariance, thus disavowing his earlier decision. Moreover, at that time he also abandoned his previous Ricci-type tensorial equations and introduced a new tensor and a new form of CCr to regain the Newtonian limit.

All this proves the great importance Einstein attributed to CCr as an expression of continuity in the development of physics.

10.5. A Correspondence Criterion for the Operational Definition of Space-Time Intervals

In order to avoid the charge that his theory had a merely mathematical import, Einstein was always very sensitive to the problem of giving a physical definition to the fundamental quantities of his equations. In his Nobel lecture,[19] he still insisted on the point that one of the relevant implications of his gravitational theory was that it comprised "only concepts and distinctions which can be associated, without ambiguity, to observable facts" and, he added, "this postulate, which pertains to epistemology, discloses something of a fundamental importance".

However, after Einstein's adoption of the new Riemanian metrics of space-time, a direct operational definition was impossible, because clocks and rods in a non-Euclidean system cannot be used as measuring devices. In the 1913 essay[20] "Entwurf einer verallgemainerten Relativitätstheorie und einer Theorie der Gravitation", Einstein and Grossman used CCr in order to give a physical definition to the new space-time interval. In order to achieve this definition, the operational definition of ds in special relativity is transferred to the correspondent ds in GR.

In order to discuss this problem, I refer here to Einstein's 1916 demonstration[21] because it proceeds along the lines of his initial 1913 demonstration originally reported in "Entwurf..".

The infinitesimal interval between two points in the space-time continuum of GR, first introduced in this essay, is:

$$ds^2 = g_{\mu\nu}(x^\alpha)dx^\mu dx^\nu \qquad \alpha, \mu, \nu = 1, 2, 3, 4 \quad , \qquad (8)$$

g_{mn} is a function of the co-ordinates x^a. Clearly, the interval above is a mathematical generalisation of the Minkowsky interval in special relativity:

$$ds^2 = \eta_{\mu\nu} dx^\mu dx^\nu \qquad \mu, \nu = 1, 2, 3, 4 \qquad (9)$$

where:

$$\eta_{\mu\nu} = diag(-1, -1, -1, c^2) \qquad (10)$$

c^2, a constant.

Since ds^2 is generally invariant in GR, there exists an infinitesimal co-ordinate transformation from an inertial local system $\{X_i\}$ to any system $\{x_a\}$, and vice versa. One is thus allowed to express the Minkowskian interval ds^2 in terms of the system's characteristic parameters $\{x_a\}$.

In fact:

$$dX_\nu = \Sigma_\sigma \alpha_{\nu\sigma} dx^\sigma \ .$$

Substituting the relation above for (9), one gets (8). The inverse transformation is supposed to hold.[22]

Let us comment upon the above passage from the conceptual viewpoint: Equation (9) has a direct operational meaning because ds^2 is the square of an infinitesimal distance between two points, measured with clocks and rods at rest in the reference system. Equation (8) does not have the same direct operational meaning because clocks and rods in a non-Euclidean reference system cannot be used as measuring devices. The reduction of (8) to (9) amounts to postulating the possibility that physical measuring objects ds do exist in general.

Thus, the measurement of x^a by rigid rods, an operation which has a physical meaning in special relativity, allows the measurement of a space-time interval ds^2 in any system through the measurement of $g_{\mu\nu}$.

This is physically possible as long as a gravitational field can be switched on and off through an appropriate co-ordinate transform, i.e., it is possible only for an infinitesimal co-ordinate transformation, but not for a finite domain where the switching operation cannot be performed in general.[23] However, the theory above implies that two infinitesimal rods of equal lengths remains equal when subjected to any gravitational field.

In summary, if an operational definition in the GR finite domain is impossible, Einstein, seemingly by *Fiat*, extended the validity of the operation from the infinitesimal to a finite space-time interval, i. e., he extended the definition valid for the Euclidean space of SR to cases in which SR does not hold, i.e., to cases in which gravitational fields cannot be switched off in general.

Einstein's procedure has deep implications for his philosophy, more precisely for his views on the requisites for the *completeness* of a physical theory. I will shortly follow these implications at some critical moments in the development of his thought.

10.6. Stratification of Theories, Physical Reality and Completeness

A conception of the progress of physical knowledge and of theory change, to which the problem of the meaning of CCr is, as we have seen, intimately related, is presented in some detail by Einstein in his 1936 essay, "Physics and reality". Scientific progress is seen as a progressive stratification of theories into different layers, each layer possessing in comparison to the lower ones an improved logical unity, consisting in a lesser number of "primary concepts and relations ".[24]

The price that must be paid in order to achieve this unity is that the upper level theories become more distant from the level of observations; this involves the risk that these theories lose contact with observations. Let us note that it is this contact which distinguishes a physical from a mathematical theory. One can, therefore, argue for the importance of CCr in its role of attributing a physical significance to the abstract entities of the higher level theory.

Precisely in order to achieve this significance the operational definition of the lower level quantity is transferred to the corresponding higher level, as I have shown in the *ds* case above. In this transfer of meaning, a correspondence criterion is raised to the status of a principle: there must be correspondence.

It can be argued that Einstein had in mind the idea of stratification when he developed his gravitational theory, comparing it with special relativity and Newton's theory of gravitation. However, it is clear that this relation of transferred meaning does not consist in a transfer of a sensory likeness, but in the possibility of establishing a term-to-term correspondence between the concepts in the theory and the perceptions at the observational level.

This special kind of relation is by Einstein expressed through a metaphor:

This relation is not analogous to that of soup to beef but rather of the cloak-room ticket to the overcoat.[25]

Concepts are related to observables as the wardrobe ticket is related to the overcoat. In as such as nothing in an overcoat in itself refers to the cloackroom ticket, this metaphor illustrates Einstein's view [26] that, once the parallelism of laws between concepts and perceptions has been established, concepts do give physical meaning to observables, not vice versa.[27]

This is consistent with Einstein's conclusion in his 1936 essay, "Physics and Reality", that concepts derive their physical meaning primarily from the whole theoretical context into which they are inserted and that only secondarily does this meaning depends on their being related to observables, e.g., through an operational definition.

CHAPTER 11

EINSTEIN'S LIFE-LONG DOUBTS ON THE PHYSICAL FOUNDATIONS OF THE GENERAL RELATIVITY AND UNIFIED FIELD THEORIES

11.1. A Foundational Problem in Einstein's Relativity

To the best of my knowledge, it can safely be argued that the majority of recent studies on the foundational difficulties of quantum physics start from the assumptions that the conceptual foundations of classical physics and relativity theory are clear and unproblematic and the present problems should concern only quantum physics. Yet, a simple inquiry into the literature and especially into Einstein's epistemological writings shows that, contrary to the view of a supposedly well-founded classical physics and relativity, important foundational problems in these sciences are still in need of further analysis.[1] The belief that the only foundational difficulties belong to quantum physics (henceforth QP), as if they alone existed against an ideal unproblematic background of classical physics (henceforth CP) and relativity theories (RR), results in a limited approach in examining the historical documents.

It is known that Einstein's epistemological views were at times misinterpreted by physicists and philosophers. I argue that these misinterpretations were in part the result of a certain amount of ambiguity [2] in Einstein's early epistemology. As an example, let us take the Bridgman case:[3] in 1949 he considered Einstein's special relativity (henceforth SR) to be the "manifesto" of the fruitfulness of operational definitions for physical quantities. In his later years Einstein rejected Bridgman's interpretation.[4] As another example of this misunderstanding, let us take the case of Werner Heisenberg. In 1926, shortly after the publication of his paper on the presumed *Anschaulichkeit* of QP, Heisenberg confided to Einstein that he had actually taken the idea of observable quantities from Einstein's RR. Einstein quickly discredited Heisenberg's interpretation saying that, just to the contrary, he maintained that it is the theory which ultimately decides what can be observed and what cannot.[5]

I will show that Einstein's approach to the problem of the meaning of the space-time interval in his SR easily lent itself to Bridgman's and Heisenberg's criticisms.

11.2. Einstein's Problems with the Stipulation of Meaning for the Riemanian Space-Time Continuum

Although Einstein frequently discussed the problem of the theory-experiment relationship, it has gone almost unnoticed by Einstein scholars that, in his papers, this problem is presented in the form of a search for criteria for attributing physical significance to the concepts of the relativity and unified-field theories, i.e., in the form of a stipulation of meaning.

Einstein first confronted this problem in his early approaches to special relativity in 1905 and, until his last years, he never ceased to search for possible solutions. Thus, problems concerning the "stipulation of meaning" enter into all of Einstein's methodological discussions on general relativity and its generalisation into the unified-field theories. In these discussions, Einstein often came in touch with the methodological views of mathematical physicists such as Weyl,

Eddington, Levi-Civita, et al. In his 1923 Gothenburg lecture he laid down explicitly a *stipulation of meaning* (SM) for the concepts of physics:[6]

Concepts and distinctions are only admissible to the extent that observable facts can be assigned to them without ambiguity.

He found that in classical mechanics (CM) the definitions of such concepts as inertial system and free body are circular; hence CM transgresses SM:

Note in passing that the logical weakness of this exposition [i. e. the exposition of CM] from the point of view of the stipulation of meaning is the lack of an experimental criterion for whether a material point is force-free or not; therefore the concept of the inertial frame remains rather problematic.[7]

Contrary to what might be expected, SR faces the same difficulty as CP. The concept of an inertial reference frame is also a fundamental foundational problem for that theory, because, from the point of view of SM, a reference frame is just a combination of rigid rods. But, as is known, rigidity would allow instantaneous signal transmission, thus contradicting a fundamental postulate of SR. This justifies Einstein's conclusion to his Gothenburg lecture:

I am mentioning these deficiencies of method because in the same sense they are also a feature of the SR in the schematic exposition which I am advocating here.

Concerning Einstein's problem with rigidity, it is worth mentioning that, in 1909 Max Born called attention to this problem and proposed[8] an original solution to it in various essays.[9]

Because SR, like CP, is not able to find a satisfactory SM-observant physical meaning for rigidity, it was logical for Einstein, in the same Gothenburg lecture, to explore GR as a possible basis for the solution of his problem. He then hinted at an indirect criterion as a guarantee for the physical meaning of the concepts: the *simplicity criterion* for theory validation.[10]

However, in spite of the above criterion, which appeared to be a way of circumventing the problem of SM, he shortly thereafter in the same paper[11] presented GR as a radical solution to the rigidity problem: the abolition of finite rigid inertial frames and their substitution with local inertial frames. In a gravitation-free space the infinitesimal space-time interval of GR:

$$ds^2 = g^{mn} dx_m dx_n ,$$

has to coincide with :

$$ds^2 = c^2 dt^2 - dx^2 - dy^2 - dz^2 ,$$

its correspondent in the pseudo-Euclidean space of SR.

Einstein initially adopted this solution, i.e., the solution via the so-called correspondence criterion or method, in his 1912 attempt to generalise SR into GR.[12] However, he was aware that this solution contradicted the foundational assumptions of GR. In his own words:

It [the solution] was inevitably fatal to the simple physical interpretation of the coordinates, because it could no longer be required that coordinates' differences should signify direct results of measurements with ideal scales or clocks.[13]

11.3. "Correspondence" as a Logically Asymmetric Method for Correlating Concepts and Perceptions. Einstein's Desire for a Purer Method

Given the premises above, it is not surprising that Einstein, in his 1936 essay,[14] "Physics and Reality", returned to the problem of an SM for *ds,* but this time through a more general approach founded on the "*Stratification of the Scientific System*". A physical theory is stratified into various levels: the lower level, which is also the most primitive in a diachronic sense, comprises concepts that are more directly related to perceptions (*Empfindungen*) and to the theorems which connect them.

Although the upper level concepts are more distant from perceptions, this defect is balanced by an advantage: the theory gains in simplicity, i.e., in the clearness and distinctiveness of their axiomatic foundation, what it loses in empirical contact. However, in order to make contact with the empirical level, the upper level concepts need to be reduced to their correspondents in the lower level. This is achieved by way of a mapping process. The mapping is achieved in the so-called Correspondence area, by relating the upper level concepts to their lower level correspondents,[15] but not vice versa. In this sense there is an asymmetry in the correspondence relationship, the correspondence is univocal, not bivocal:

The relation is not analogous to that of a soup to a beef but rather of a cloakroom ticket to an overcoat.[16]

Once the correspondence is established, the upper level concepts receive their physical meaning from their corresponding lower level concepts.[17] In his view, Einstein confirms and generalises his correspondence criterion of 1912 into a general feature of theories.

If one concedes that the above stratification and the related hierarchy of concepts somehow absolve the upper level concepts from their transgression of SM, one should also admit that, concerning SM, this new method also amounts to a transfer of meaning from the lower to the higher level through correspondence rules which possess many degrees of freedom. For this reason, this transfer has sometimes been nicknamed "a transfer of meaning by decree".[18] It risks endangering the physical foundation of the higher level theory by depriving it of its autonomy with respect to its lower level counterpart. In this sense it appears a hybrid method.

In 1923 Einstein mentioned[19] an alternative method, which he considered to be *purer* (supposedly, in the sense of being less hybrid). He attributed it to the theories of Levi-Civita, Weyl and Eddington. In these discussions, Einstein often came in touch with the methodological views of these mathematical physicists, all interested in various formulations of the unified-field theories (hereafter UFT).

As is known, Levi-Civita introduced[20] the notion of parallel displacement into differential geometry in 1917. Through this contribution, he provided what seemed to be an indispensible tool for casting GR into a coordinate-free geometrical form, thus overcoming Einstein's problem with the rigid rod for the coordinates' physical definition. The Levi-Civita theory had a great and immediate impact on Weyl's influential *Raum, Zeit, Materie* [21] of 1918. Weyl's theory made a considerable impression upon theoreticians and on Einstein himself, who wrote that its depth and boldness must charm every reader.[22] In Weyl's method, the meaning of concepts in a theory at level B > A should be founded without any recourse to the corresponding concept at level A. In 1918-19, the recourse to A was avoided by Levi-Civita, and followed by Weyl through the choice of an *affine geometry*. As is known, this geometry assumed that *parallel transport*, i.e., the parallel displacement of a vector, is accompanied not only by a change in the vector orientation, as in Riemanian geometry, but also by a change in the vector's length.

Therefore it is not surprising that Einstein took this theory as an example of a *purer theory* , not committed to a more or less direct operational definition of the coordinates.

In 1923, Einstein introduced the new argument for a *purer method* after complaining that it is methodologically unjustifiable to base all physical considerations on the rigid or solid body and then to reconstruct that body atomically by means of elementary physical laws which in turn have been determined by means of the rigid measuring body.23

This argument continues in Eintein's new requirement for a *complete physical theory*: *the over determination of physical equations*. In fact, the above passage expands upon the features of the *purer method* which supposedly would have avoided the transgression of SM:

Certainly it would be logically more correct to begin with the whole of the"laws and to apply the "stipulation of meaning> to the whole first, i. e., to put the unambiguous relation to the world of experience last *instead of already fulfilling it in an imperfect form for an artificially isolated part, namely the space-time metric*. At the close of our considerations we will see that in the most recent studies there is an attempt, based on the ideas of Levi-Civita, Weyl, and Eddington, to implement that *logically purer method* [italics are mine].

I will show that the problems concerning SM were also present in Einstein's generalisation of GR into a UFT.

11.4. Over-Determined Theories Avoid the SM Transgression

Einstein sketchily drew up in his 1936 "Physics and Reality" a possible mode of meeting the requirement for an over-determined theory. He exemplified how a field theory can account for particles in the form of a Schwarzschild-type singularity-free solution for a modified differential equation: $g^2 R_{ik} = 0$, in place of the former equation for empty space, $R_{ik} = 0$. Einstein mentioned this example in connection with the purer theories of Levi-Civita, Weyl and Eddington, thus presenting it as a return to his 1923 proposal above for an over-determined theory.[24]

In 1945, Einstein again took up his 1925 asymmetric theory, remaining faithful to this approach until the end of his life[25]. As is known, a remarkable feature of this approach is the non-linearity of the resulting electromagnetic equations. It is this non-linearity which fulfils Einstein's request for over-determined field equations whose spherical-symmetric singularity-free solutions can be interpreted as elementary particles. It can be reasonably argued that the complete fulfilment of the requirement above would correspond to Einstein's ideal of a *purer theory*.

It is well known that this ideal theory was never achieved during Einstein's life and that it was considered hardly realisable by the majority of physicists when he died in 1955; the more so after the expansion of the QM approach to particles in modern theory.

Given the premises above, it can be argued that Einstein's 1949 discussion with Hans Reichenbach has to be understood as a further clarification of his 1923 and 1936 search for a *purer method*. Einstein's last "discussion" with Reichenbach was presented in Einstein's 1949 "Reply to Criticism",[26] a part of his contributions to Schilpp's memorial work *Albert Einstein: Philosopher Scientist*. It offers an example of Einstein's mature thought on the difficulties of the meaning problem in relativity and UFT.

Einstein's aim was to refute Reichenbach's argument on the possibility of deciding what is the real geometry of the world through an experimental check on the Euclidean congruence of physical rods.[27] Einstein's refutation runs as follows: in order to check an Euclidean congruence one needs rigid rods, but to control rigidity one actually has to resort to physical laws such as those for the control of temperature-constancy, elasticity-coefficient, etc. laws that, in their turn, require a prior assumption of rigidity for their foundation. In short, Reichenbach's proof of rigidity is considered by Einstein logically circular. (It is worth noting that Einstein's charge of circularity is, at bottom, an argument that Einstein also presented in his 1923 search for a SM-observant definition of rigidity)."

In 1949, Einstein's own thesis is that "Meaning" can be attributed to the individual concepts and assertions of a physical theory and to the entire system only insofar as it makes what is given in experience "intelligible""[28].

It is evident that this thesis echoes the famous metaphor of the Kantian Copernican revolution, i.e., Kant's assertion that rationality is a precondition for reality and not vice versa. In fact, in an ensuing passage, Einstein admits his late adherence to the basic tenets of Kant's philosophy.[29]

11.5. Einstein's Final View: the Incompleteness of General Relativity Seen as Lack of a Satisfactory Criterion for the Postulation of Meaning

The Einsteinian method of giving meaning through CR transfer did not satisfy his own requirements for theory in 1949. His mature reactions to this method are expressed in this passage:[30]

One is struck [by the fact] that the theory (except for the four-dimensional space)[31] introduces two kinds of physical things, i.e., (1) measuring rods and clocks, (2) all other things, e.g., the electro-magnetic field, the material point, etc. *This, in a certain sense is inconsistent*; strictly speaking measuring rods and clocks would have to be represented as solutions of the basic equations (objects consisting of moving atomic configurations), not, as it were, as theoretically self-sufficient entities. However, the procedure justifies itself because it was clear from the very beginning that *the postulates of the theory are not strong enough to deduce from them sufficiently complete equations for physical events* sufficiently free from arbitrariness, in order to base upon such a foundation a theory of measuring rods and clocks [Italics added].

Notice that in this passage the 1923 SM problem the introduction into theory of measuring rods and clocks is joined to the 1936 under-determination problem (the postulates of the theory are not strong enough). I interpret this fact as evidence that, in 1949, the mature Einstein believed that the two problems were parts of the more general problem involving the physical basis of GR and UFT.

In fact, in the same year, he criticised the current theory of relativity (i.e. GR) for not meeting the requirements[32] for a physical meaning of *ds*:

For the construction of the present theory of relativity the following is essential:
(1) Physical things are described by continuous functions, field-variables of four co-ordinates. As long as the topological connection is preserved, these latter can be freely chosen.
(2) The field-variables are tensor components; among the tensors is a symmetrical tensor g_{ik} for the description of the gravitational field.
(3) There are physical objects, which (in the macroscopic field) measure the invariant *ds*.
If (1) and (2) are accepted, (3) is plausible, but not necessary. The construction of mathematical theory rests exclusively upon (1) and (2).
A *complete* theory of physics as a totality, in accordance with (1) and (2) does not yet exist. If it did exist, there would be no room for the supposition (3). For the objects used as tools for measurement do not lead an independent existence alongside of the objects implicated by the field-equations[talics added].

Here, for the first time, the completeness theme is connected with that of an over-determined theory. By relating the latter statement on completeness to the above passage,[33] one is led to the following definition: a theory is said to be complete if *the postulates of the theory are strong enough to deduce from them sufficiently complete equations for physical events sufficiently free from arbitrariness*. A physical theory of GR, i. e., a theory that gives physical meaning to the concepts of measuring rods and clocks by representing them as solutions of the basic equations, has to be based on postulates of the above stronger type. The impossibility of deducing from the foundational postulates or axioms of GR a physical meaning for *ds qualifies GR as an incomplete theory.*

The above definition neither excludes nor contradicts other interpretations of Einstein's views on incompleteness, such as those presented in several recent valuable studies.[34]

I present it as a definition which conforms to Einstein's 1949 position.

An interesting viewpoint is provided by Einstein's consideration that, due to the theory's *incompleteness,* postulates 1) and 2) are for the time being only able to characterize a mathematical and not a physical theory. This consideration implies that a future ideal *complete theory* should also represent a synthesis between pure mathematics and physics. In Einstein's ideal, this synthesis would consequently abolish the distinction between mathematical-physics and physics, the two traditions which were often counterposed in the historical development of Western physics. This *type of synthesis* was not accomplished by Einstein himself nor by subsequent quantum physicists. They simply took another direction, somehow bypassing, rather than solving, the problems that Einstein had explored.[35]

I do not consider Einstein's failure in reaching this synthesis as a shortcoming of his science, a missing exhaustivity, as it were, of his methodological discourse. Such a reductive view would, I believe, go against, among other things, the largely aknowledged value of his theories. Rather, I believe that by always stating forthrightly his methodological doubts Einstein made an important contribution to our understanding of the philosophical implications of theoretical physics. In this connection, I would like to refer to a passage by Gerald Holton:

By always stating forthrightly and with eloquence his redefined position, Einstein not only helped us to define our own, but also gave us a virtually unique case study of the interaction of science and epistemology. [36]

In support of his thesis, Holton quotes Max Planck's statement which surprisingly confirms the above view:

A science is never in a position completely and exhaustively to solve the problem it has to face. We must accept that as a hard and fast, irrefutable fact, and this fact cannot be removed by a theory which restricts the scope of science at its very start.[37]

It is noteworthy that Planck's statement generalises Einstein's view on the inherent incompleteness of his own theories.[38]

11.6. Conclusions

It has been rightly remarked[39] that the scientific program underlying Weyl's theory and the companion geometrical UFTs were closely connected with the Göttingen tradition in mathematical physics. It represented a new form of interaction between mathematics and physics that was characteristic of the non-classical theories of the twentieth century. Until the emergence of quantum physics (QP), the great theoretical penetration and the mathematical perfection of the geometric UFT program were seen as genuine advantages, notwithstanding their exceedingly weak connection with experience.[40]

However, the UFT program increasingly brought to light, not only the great heuristic possibility of its new form of connection between physics and mathematics, but also certain dangers and difficulties, such as an overemphasis on the role of mathematical structure and an underevaluation of the experimental and empirical aspects of theory.[41] The latter difficulties should have appeared as grave defects especially when compared to the theoretical eclecticism of the QP program, and the predominance within it of the empirical over the theoretical.[42]

Clearly, Einstein's completeness requirement is also related to the Bohr-Einstein controversy concerning the foundation of QP. Recent studies have interpreted this controversy in the context of the known split between two schools of post-Kantian philosophy.[43] In these studies, Einstein's and Bohr's different views on the foundational problems of modern physics are seen as a contrast between the two trends in post-Kantian philosophy, usually paraphrased in the keywords: *Anschauung* and *Symbol*.[44] In short, in Bohr's view the formalism of QP represented a purely symbolic scheme that could not be directly visualisable. In contrast, the classical and the Einsteinian formal approach aimed at a direct intuitive interpretation, i.e., this approach aimed to be *anschaulich* in the classical Kantian sense.[45]

In contrast with the views stated above, I wish to mention an interesting thesis by Arthur Miller, one which intends to point out an element of essential continuity between methodologies underlying both UFT and QP.

In his treatment of the theme of "visual imagery",[46] to which Miller has devoted outstanding publications, he distinguishes between two different processes of adaptation of visual imagery in the growth of physics, one in a form of *visualisability (Vb)*, more keen to the perceptive level of our representations and predominant in classical physics, and the other, *i.e. visualisation (Vs)*, which became indispensable in order "to explore worlds beyond our sense perception, such as the atomic realm".

In the author's view, the transition from *Vb* to *Vs*, distinguishes modern from classical physics, and it was primarily ruled by the new status that mathematics has conquered in modern physics.[47] Consequently, the *Vb* —> *Vs* transition, radical as it may seem, does not seem to represent a fracture in the scientific tradition, mainly because two powerful factors of abstraction have been always at work, i.e., mathematics and metaphorical transformation of language.

After reading Miller's essay, one can readily identify *Vb* with Einstein's *Anschaulichkeit*. In consequence, in this view the transition from UFT to QP would not present any element of discontinuity. Clearly, my position above is in contrast with Miller's view because I find that the ideal of a convergence of math into physics, sought in vain in the mathematical-physics tradition, was not achieved in Einstein's relativity and UFT. QP somehow bypassed the difficulty.

The fact that the convergence was not reached by the classical theoreticians nor by quantum physicists, opens a new area for historical and critical research on the foundational problem[48] of both GR and QP.

CHAPTER 12

CORRESPONDENCE AND COMPLEMENTARITY IN NIELS BOHR'S PAPERS: 1925-1927

12.1. Karl Popper's Criticism of Bohr's Complementarity

Niels Bohr devoted many of his reflections to epistemological and philosophical matters, though he declared that he did not consider himself a philosopher. The philosophical sides of his work were differently evaluated by scientists and philosophers: highly estimated by scientists (Heisenberg, Rosenfeld, Weizsacker, etc.), severely criticised[1] by epistemologists such as Popper, Margenau, Park, etc.

In any case, Bohr presents problems to philosophers and scientists. I think that the case of the philosophical relevance of Bohr's contributions cannot be easily set aside. I have selected here as my specific topic the problem of the relation between correspodence (Cr) and complementarity (Cm) in Bohr's work from 1925 to 1927.

It is well known that the two principles have distinct origins in the development of Bohr's thought and that their bearings on the development of quantum theory (QT) are also varied. Karl Popper was perhaps one of the first among the philosophers to underline this difference. While he evaluated the fecundity of Cr for the development of QT, he denied[2] any physical and philosophical relevance to Cm. This discredit is for him part of the philosophical and scientific irrelevance that he attributes in general to the so-called Copenhagen School.

Popper's argument has prompted both praise and criticism- from historians and philosophers.[3] In this chapter I will argue whether Cr and Cm are to be considered conceptually independent positions or whether Cm represents just a consistent expansion of Cr. If the second alternative is supported by the historical evidence, Popper's argument loses much of its strength, to say the least. On this occasion I will take into account a short interval of time in the span of Bohr's scientific life, the years 1925-1927 when the fundamental ideas of the Cm theme were laid down.

My method of inquiry consists in the selection and analysis of the main and more relevant themes of Bohr's discourse and in the tracing of their permanence or of their modification throughout the development of Bohr's thought.

I am not alone in remarking[4] that the thematic development, in Bohr's writings, is characterised by an initial vagueness, which, however, does not preclude its essential correctness. Often the meaning of his statements is not fully formulated at the beginning, but it improves in their subsequent formulations, progressing towards more precise definitions. In the majority of cases the future developments justify the present ones.[5] This implies that Bohr's thought is in a fluent state and that the meaning of some of his concepts can be better clarified only through an examination of the whole context of the author's work. In view of these remarks I consider the historical approach the more appropriate methodology for Bohr's case.

The volumes of Bohr's *Collected Works,*[6] a rich assembly of documents, letters, events present the historian of science with a very helpful source for the reconstruction of the exceptional climate of those days, when the fundamental ideas of modern physics were brought to light.

12.2 Bohr's Correspondence Principle in 1925

According to W. Krajewski, the basic ideas of Cr are contained already in Bohr's 1913 trilogy,[7] in which he formulated the quantum postulates concerning the orbits of the electrons in the atom. Bohr did not use the term "correspondence" in this period, but the term "analogy" between quantum and classical laws.

The expression "correspondence" (Cr) appeared for the first time in Bohr's 1920 "Über die Linienspektren der Elemente", and the "Cr principle" (CrP) in an Academic memoir,[8] shortly afterwards. In the same memoir Bohr further extended Cr, interpreting Fourier's expansion coefficients by Einstein's "probability factors of spontaneous emission".

Kramers (1919) had applied with surprising success Bohr's formula to the calculation of the relative intensities of the components of the fine structure in the Stark effect. But this many-sided extension of the CrP was in danger of making it appear[9] as "a magic wand that allowed the results of the classical wave theory to be of use for the QT". Hence, the need arose for a reinterpretation of Cr in 1925. The first but very significant approach to the problem of the correlation between Cm and Cr is in Bohr's 1925 paper "Atomic Theory and Mechanics".[10] One of the central themes in this paper concerns the "formal nature of the frequency condition".[11] The definition of the photon's frequency is considered to be related to the classical theory (CT) of the wave-phenomenon of interference and, as such, it is deemed to contradict the presumed photon's particulate nature.[12] According to this criticism, Bohr does not accept the concept of the photon as a particle this is one of his points of controversy with Einstein.

In order to avoid this contradiction in his 1925 paper, Bohr could have chosen to reject completely visualisation (a kind of generalisation of his previous rejection of the possibility of visualising photons) and accept RC in its formal aspects (after all, this proposal had been accepted in the past in the form of a Kirchhoff-type mathematical phenomenology). However, while this could have been Dirac's proposal at that time, it could not have been Bohr's: a total rejection of visualisation is not Bohr's way. His choice at this time (1925) is instead: a "restriction" of visualisation.

His "renunciation" in 1925 of any attempt to form a "mechanical picture"[13] of the photon (i.e., his refusal to picture it as a particle), arose from his belief in the fact that "mechanical pictures" of the phenomena of the old quantum theory should not concern most of the new physics, only those few parts in which quantum laws agree (formally) with classical laws.[14]

He presents many reasons in favour of his choice: one is the success of visualisation in CT. In fact, Bohr introduced his 1925 paper by enlisting a long series of successes of classical theories in explaining atomic phenomena: the statistical theory of fluctuation, Rayleigh's scattering of blue light, Lorentz's explanation of the Zeeman effect, the discovery of the electron, etc. Quantum theory, for those aspects which were included under Cr, is then also presented by Bohr in a role in which they confirm the validity of visualisation in classical physics.

For example, after mentioning difficulties connected with the "quantum conditions", which seemingly might suggest the "abandonment of a space-time description", he adds:

Nevertheless, it has been possible to construct mechanical pictures of the stationary states which rest on the concept of the nuclear atom and have been essential in interpreting the specific properties of the elements.[15]

Notice that the space-time description is here considered as a prerequisite for visualisation.

In the same page he affirms:

Nevertheless, the visualisation of the stationary states by mechanical pictures has brought to light a far-reaching analogy between the quantum theory and the mechanical theory.

The fact is that CT represented for Bohr a glorious patrimony of science which had to be recovered and salvaged in some way. Soon this exigency will be partially fulfilled by complementarity but, for the moment, he salvages visualisation and CT. Consequently, he refers to "the asymptotic agreement between spectrum and motion". This includes, of course the well known derivation of the Rydberg constant in terms of classical laws. Moreover, Bohr mentions the extension of Cr not just to frequency, but also to amplitudes in the well known Fourier series expansion of the multiperiodic atom.[16] At the end of his list of successes of Cr, he also refers to the so-called selection rules, a consequence of the development of Cr in the theory of Sommerfeld phase integrals and in Ehrenfest adiabatic invariance.[17] He remarks with special emphasis that the latter use of Cr allowed the prediction of the intensity of the lines in the Stark effect.[18]

Nevertheless, in his essay Bohr counterpoises the successes of Cr with a list of "profound difficulties" in the classical explanation (i.e., in the forming of mechanical pictures) of quantum theories. The list opens with Planck's well known difficulty in introducing his quantum of action, soon followed by difficulties with the anomalous Zeeman effect, the Rutherford model of the atom, Bohr's own theory of stationary states, and in the end with Compton and Simon's interpretation of the Compton effect.[19]

Continuing the above list of difficulties with visualisation, Bohr devotes much attention to the difficulties of the Kramers-Heisenberg theory of optical dispersion:

> While this description of optical phenomena was entirely in harmony with the fundamental ideas of the quantum theories, it soon appeared that it stood in *strange contradiction* to the use of the mechanical picture previously employed for an analysis of the stationary states [my italics].

When he wrote the paper, Bohr was still impressed with the disturbing feeling arising from the failure of the Bohr-Kramers-Slater theory. His initial great expectation for Cr and his later disappointment are evident in a letter to Pauli. He was at first "impressed by the surprisingly wide applicability of mechanical pictures".

"For this reason", he and Kramers tried to solve "the problem of the hydrogen atom in crossed electric and magnetic fields" with methods "outside the region of multiple periodic systems. When this belief turned out to be wrong, it became clear that a classification of the stationary states, based on mechanical pictures, could not be carried out".[20]

At this stage it seemed to Bohr that he was faced "not with a modification of classical laws", but with "an essential failure of the picture in space and time".

Confronted with this failure, Bohr adopted the following solution: he reinstated Cr in the role of a principle, a guide for the further development of QT, a correspondence principle (CrP). But in this role, Cr needed "rationalisation". It needed to be removed from its empirical state, and to be promoted to the role of a method for "transforming" classical laws into quantum laws:

The CrP expresses the tendency to utilise in the systematic development of the quantum theory every feature of the classical theories in a *rational transcription* appropriate to the fundamental contrast between the postulate [i.e. the quantum postulate] and the classical theories [my underlining].[21]

This "rational transcription" could be saved from "contradictions" and accepted in the new physics only at the cost of introducing a criterion for an appropriate "restriction" or, equivalently, "limitation" in picturing physical objects.

The introduction to Cr in "Atomic Theory and Mechanics", begins with the term *nevertheless*.. This word suggests that Cr is a way of escaping the fundamental contrast by accepting limitation. The word *"nevertheless"* is again repeated a few passages below.[22]

Bohr mentions "the fundamental difficulties involved in the construction of pictures of the interaction between atoms either by means of radiation or by collision". To solve these difficulties what seems necessary is the "abandoning of mechanical models in space and time which is so characteristic a feature of the new quantum mechanics". In fact, only if "any mention of the time at which transitions take place" is "definitely" avoided, can the new theory represent the transition. Bohr considers that this "restriction"... is typical of the attack on the problem of the constitution of atoms based on the postulates of the QT".

On the other hand only this restriction "allows some aspects of the analogy between the QT and the CT to come to light".[23]

In 1925, the criterion for introducing this restriction in CrP is the one offered by Heisenberg's theory of matrix mechanics.[24] As is known, in the same year Heisenberg presented this paper in the context of a "theory of observables". In his article Heisenberg generalised Bohr's rule for frequencies and extended it to amplitudes. Heisenberg called his proposed method "a method of determining quantum-theoretic data using relations between observable quantities", thus offering to Bohr the key for his new interpretation of Cr as a rule for a "rational transcription" restricted to observable quantities.[25]

Bohr's special interest in the "observables" theme in Heisenberg's paper is, in my opinion, related to his initial awareness of the idea of Cm, as hinted at by his enthusiastic comment on the whole of Heisenberg's paper: "The whole apparatus of quantum mechanics can be regarded as a precise formulation of the tendencies embodied in the correspondence principle".[26]

In conclusion, the restriction that Bohr accepted in 1925 in order to avoid the "strange contradiction found strong support in the fact that those entities introduced the "contradiction", such as frequency of revolution, trajectories, etc., "are not open to comparison with observation"[27], i.e., are not "anschaulich" in Heisenberg's sense.

Thus, in consequence of Heisenberg's Matrix Theory, Bohr's Cr underwent in 1925 an important shift of meaning.[28] The criterion for his restriction consisted then in the "transcription" of just those physical properties such as frequencies and amplitudes that are "observables". In this sense the transcription is a "rational transcription".

12.3. From the 1925 "Restriction" to the 1927 "Limitation": Complementarity

The type of "restriction" Bohr introduced in 1925, following Heisenberg's matrix theory, will be modified in 1927, as a consequence of the appearance of Heisenberg's *indeterminacy relations* (IR). I think that the 1927 "limitation" Bohr introduced through his complementarity (Cm) is a natural development, due to the notion of IR, of his 1925 "restriction". Among others, this passage from a letter from Bohr to Einstein (dated April 17, 1927) confirms the point:

This very circumstance that the *limitation* of our concepts coincide so closely with the *limitation* on our possibilities of observation, permits us as Heisenberg emphasises to *avoid contradictions* [my italics].[29]

The shift from the 1925 "restriction" to the 1927 "limitation" is especially evident in some of the papers that Bohr devoted to Cm in 1927 and 1928 particularly in his "The Quantum Postulate and the Recent Development of Atomic Theory", a 1927 typewritten copy found among Darwin's papers.[30]

The reason why I find this paper significant for my thesis is that in it Bohr expresses his ideas in a "burgeoning state", under the intellectual shock of the publication of Heisenberg's famous 1927 article.[31]

Let us carefully think over the introductory statement:[32] the Quantum Postulate is introducing "fundamental limitations in the classical physical ideas, when applied to atomic phenomena" while, on the other hand, "the interpretation of the experimental material rests extensively upon the classical ideas". The first part of the statement is a new interpretation of the by then secure 1925 "limitation" theme, and the second passage announces Cm.

The passage connects "limitation" with the emerging theme, the usefulness not yet indispensability[33] of classical ideas for the interpretation of experiments. However the usefulness of classical ideas will soon develop in that they become indispensable for the "description of experience". In fact a few passages below Bohr explicitly declares:

[...] radiation in empty space as well as isolated material particles are *abstractions*, their properties on the quantum theory being observable and definable only through their interaction with other systems. *Nevertheless* in the present state of science these abstractions are *unavoidable* for a description of experience [my italics].

I believe that, in Bohr's mind, the statement that the classical concepts are "abstractions" should be related to his previously statement on the impossibility of using *classically* "radiation in empty space as well as isolated material particles". In fact in 1925 and 1926 he remarked on the impossibility of visualising, i.e., using classically, the rotating electron. The quantum postulate confers to this concept a symbolic[34] character:

The symbolic character of these pictures can scarcely be more strongly emphasised than by the fact that in the normal state no radiation is emitted, although according to the mechanical picture the electron is still moving.[35]

The above is a preview of the complementarity theme, in as such as Cr will soon be called on to perform a new function: "the more general need to speak of quantum events with the aid of classical (every day) concepts". However, this need can be only satisfied in as much as these concepts are used symbolically. This condition will later become[36] "a condition for unambiguous communication".

In the introductory chapter of the 1927 typewritten paper, a third main theme deals with the complementarity relationship between "observation and definition": if the definition (i.e., space-time description) of a system of objects claims to eliminate external disturbances, then, conversely, no observation is possible without an interaction which disturbs the objects, thus rendering impossible its definition. The term "definition", in Bohr's view, should then be understood not in the sense of a verbal definition but as the process of obtaining a space-time (ST) description of events (i.e., a definite representation). In the ensuing passages Bohr shows how this fundamental conception applies to the case of the interaction between light and matter.

According to Kalckar:[37] "The first reference to the notion of complementarity is probably that found in a manuscript of July 10, [1927]...[an] extremely interesting document".

Here is the reference:

> The theory exhibited duality when one considered on the one hand the superposition principle and on the other hand the conservation of energy and momentum. [...] Complementary aspects of experience that cannot be united into a space-time picture based on the classical theories.

At this point, let us briefly trace the evolution of complementarity in Bohr's papers of 1925-27. The initial complementary pairing, i.e., Superposition Principle-Conservation Laws, evolves towards the Definition- Observation pairing and the ST description-causality pairing. The latter is at the end transformed[38] into the wave-particle pairing and its physical illustration: on the one hand the validity of the wave-theory in interference phenomena in empty space (and in the classical optical properties of a material medium), and, on the other "the laws of conservation of energy and momentum for the interaction between radiation and matter".

It is worth making the following remark in conclusion: in 1927 Bohr presented Cm through arguments on a general epistemological level without any connection with Heisenberg's deduction and interpretation of IR. Heisenberg's interpretation is merely presented as one of the Cm's physical illustrations.

12. 4. Bohr's Deduction of IR

Although IR is initially formulated by Heisenberg, we owe to Bohr an original derivation of its formula. Bohr starts from Einstein's relations:

$$E = h\nu \qquad P = h\sigma.$$

These, according to Bohr, only connect the wave-like features of physical reality, frequency ν and wave number σ, with its "particulate" features, energy E and momentum P. While in classical physics, the two features are said to represent distinct domains of quantities, in quantum theory the two domains are inter-mingled, being connected through the above stated relations. The quantum of action h simply correlates the first to the second domain.

Therefore in quantum mechanics (QM), it is as if the above relations have a strange property of "wholeness", which seems to deny the possibility of a distinction between ondulatory properties in one phenomenon and particle properties in another (a possible distinction in classical physics). This same "wholeness" was seen in 1925 under the theme of "the fundamental contradiction". In Bohr's new outlook, these distinct pictures are "abstractions", which, however, "in the present state of science ... are unavoidable in a description of experience".

In the frame-work of Bohr's new outlook on the derivation of IR, if one wishes to rescue in the wave picture, its "particulate" aspect ("[...] in order to represent individuals", not abstractions), one should use wave packets and identify the velocity of the particle with the group velocity of the associated wave, confining the wave-train to a limited extension and to a limited time of observation.

We then have:

$$\Delta t\, \Delta x \sim 1 \qquad \Delta x\, \Delta \sigma \sim 1$$

(due to the limitation in frequency and space spread of the associated De Broglie wave).

By inserting Einstein's relations (above) into the equation, Bohr obtains his IR:

$$\Delta t\, \Delta E \sim \Delta x\, \Delta P_x \sim h$$

P_x : x component of momentum P.

I believe that Bohr's special procedure of deriving IR deserves special attention. Very appropriately, Max Jammer underlines the fact that Bohr deduces these relations by way of associating a particle with a De Broglie wave-packet an alien procedure to Heisenberg at that time and that "Bohr's derivation of the IR differs fundamentally from Heisenberg's derivation". The matter of controversy, however, "was not about the conclusions, that is, about the validity of the IR, but rather about the conceptual foundations on which they were established".[39]

The bitterness of the controversy in spring 1927 between Bohr and Heisenberg, the two fathers of QM, is an indication that their discussions involved fundamental problems the foundation of the theory.[40]

I find it remarkable that the theme of the quantum's finitude does not directly appear in the passages from whence Bohr's IR are deduced. This theme is only mentioned[41] in the introductory sentences concerning the definition-observation antinomy, but not in the deduction of Bohr's IR. Bohr's IR are, then, a consequence of our need to continue to visualise physical entities, even in QM, (the wave packet, in place of a discrete particle) thus overcoming the "fundamental contradiction".

Notice that, according to Bohr, the IR relations are "not an expression of the discontinuous change of energy and momentum say during an interaction between radiation and material particles", as in Heisenberg's interpretation. Rather they are a consequence "of the impossibility of defining rigorously such a change, when the space-time co-ordination of the individual is also considered". At bottom, they are not a consequence of introducing "particulate" features within a field theory (i.e., features related to the concept of a particle), as in Heisenberg's derivation. *Bohr's IR are the formal (mathematical) expression of Bohr's basic tenet: definition is complementary to observation.*

Even in this first deduction of IR, Bohr's intention is to emphasise that their physical meaning is in terms of the complementarity of the definiability-observability pairing, not in terms of a disturbance due to the finitude of the quantum of action, in Heisenberg's fashion. Accordingly, Cm is for Bohr the new kind of proper limitation for a consistent representation of reality in quantum theory. *While in classical theory one tries to describe the physical reality by using bot h waves and particles, because one uses abstractions,* in QM, by the symbolic method, one reinstates a more "realistic" as opposed to "abstract" approach to the problem of physical reality.

In view of the fact that, for the attribution of their meaning in the context of QM, the conceptual foundation "*à la* Bohr" of IR is as essential as their mathematical demonstration, I think there is sufficient reason to call the relations discussed above Bohr's IR, thus distinguishing them from Heisenberg's IR.

12.5. Bohr's IR in the Context of the "COMO" Congress

The former conception of Cm is consistently developed by Bohr in the various papers which are likely drafts of the Como Lecture (September 16, 1927). I will consider in particular "The Quantum Postulate and the Recent Development of Atomic Theory". This so-called second version,[42] represents a more mature elaboration of the initial Como address.

Bohr's IR are here counterpoised with Heisenberg's interpretation of his own IR in a very distinct fashion. I will not reproduce in detail Bohr's elaborate considerations on this matter, but simply underline some of the more relevant themes in the paper.

Here is one:

Before we enter upon his [i.e. Heisenberg's IR] results it will be advantageous to show how the complementary nature of the description appearing in this uncertainty is unavoidable *already* in the *analysis* of the most elementary concepts employed in interpreting experience [my italics].

Although Bohr gives Heisenberg's IR a prominent place in his analysis, this introductory passage clearly reveals Bohr's intention of distinguishing his approach to IR from Heisenberg's. One would rather talk of Bohr's personal handling of what Heisenberg had found independently from him.

Bohr continues:

In the discussion of this question it must be kept in mind that, according to the view taken above, radiation in free space as well as isolated material particles are abstractions, their properties in the quantum theories being definable and observable only through their interaction with other systems

Here Bohr initiates another fundamental theme for the foundation of his Cm: the observation of a system will always be subject to interference by the observing instruments. Conversely in every operational definition of a quantity, reference to a given instrument is unavoidable.

In the same paper, Bohr presents the demonstration of IR mainly in the fashion of his 1927 paper (the one we have examined) but for one remarkable exception: this time, his comment on Heisenberg's interpretation of IR includes a criticism of Heisenberg's definition of the operational meaning of his IR.

Bohr shows that the microscope, as such, has a decisive role in introducing an uncertainty in the transversal component of the particle's momentum, because "the finite values of the aperture"introduce an uncertainty that satisfies Heisenberg's IR. However, Heisenberg's flaw consists, according to Bohr, in having forgotten that observing localisation is not an abstract procedure, but rather that it should be defined through a real system of observation, an instrument. Heisenberg's deduction of the uncertainty in the transversal electron momentum through Doppler's and Compton's effects is thus of marginal use for demonstrating IR.[43] His operational definition of momentum through a measurement must be implemented, according to Bohr, by an operation which takes into account the measuring instrument, not just the physical process per se.

Consistent with his own premises, Bohr criticises Heisenberg's considerations on the grounds that Heisenberg does not exclude in *principle* the possibility that the conjugate variables possess accurate values, but, as a matter of fact, i.e., "a posteriori", Heisenberg denies that these values are in fact accurately measurable because of the discontinuous changes due to the finitude of h.

On the contrary, Bohr's approach to IR excludes in *principle* the possibility of measuring accurate values because of the complementary nature of description. Consequently, his approach reinforces his former foundational position, namely, the irrelevance for his IR for the discontinuity of to h :

A discontinuous change of energy and momentum during observation could not prevent us from *ascribing* accurate values to the space-time co-ordinates, as well to the momentum-energy components before and after the process" [my italics].[44]

It is not a matter of disturbed measurement but a matter of definition: in Bohr's conception, position and momentum are not indeterminate because they are disturbed by measurement, rather they cannot be sharply defined even in the absence of any observation because this is in the nature of QM objects, i.e., in the form of our knowledge of them:

The reciprocal uncertainty is essentially an outcome of the limited accuracy with which changes in the energy and momentum can be *defined* when the wave-fields used for the determination of the ST co-ordinates of the particles are sufficiently small. [my italics].[45]

What Bohr is apparently trying to say is that position is not disturbed by observation, but rather there is an innate uncertainty in the real procedure of defining localisation due to the presence of Einstein's relations.

At this point in Bohr's paper, we are presented with the main theme of his Cm: in any (operational) definition of theoretical concepts, the definition should be pursued until account is taken of the observing instrument. Bohr's former criticism of classical description is here developed towards the important theme of the inseparable link between observed system and observing instrument, a theme which Bohr will expand on[46] in the following years especially in his debate with Einstein at the Solvay meetings in 1927 and 1930.

12.6. Bohr's Irreducible Disparity between Quantum and Classical Theories

On the problem of a differentiation between Bohr's and Heisenberg's conceptions in 1927, I find that Jammer[47] expresses very clearly what is at issue:

> In Heisenberg's account of his gamma-ray microscope experiment, the existence of the electron's momentum must be assumed to exist, for otherwise it could not be disturbed, whereas the very same account intends to prove its non existence since it cannot be precisely measured.

In general, the same charge of contradiction can be extended to any sort of operational definitions which "tacitly assume to some extent the ontology of classical physics, which they explicitly purport to deny".[48] In his critique, Popper confounds[49] Bohr's and Heisenberg's positions in the Copenhagen School with Bohr's Cm interpretation. In Popper's view, they are both immersed in "the great quantum muddle". I have shown that this view is not completely supported by the historical evidence, at least as far as Bohr's work of 1927 on Cm are concerned.

If Cr and Cm are linked in Bohr's paper, any argument advanced by Popper[50] and others about the different roles and epistemological relevance of Bohr's two principles should be carefully reconsidered.

In conclusion, I think I have shown that Bohr's conceptions in 1927 avoid the contradictions inherent in Heisenberg's 1927 position[51], and present[52] a valid perspective on the foundations of QT. Both Cr and Cm are to be intended as different kinds of limitation in the use of classical conceptions in the interpretation of QT. Concerning the problem of the epistemological status of Cr, Bohr does not consider classical physics (CP) as a limit-theory of quantum physics (QP), or, equivalently, he does not believe that Cr is to be intended in the sense that CP is included within QP. Let us call Cr(1) this limit-theory or inclusive interpretation of Cr. The Cr(1) interpretation is certainly not one shared by Bohr in 1925 or in 1927. M. Jammer rightly remarks that Cm would contradict Cr(1):

Later Bohr's conception of Cm, which gave a deeper significance to his previous ideas on the irreconcilable disparity between classical and quantum theory, precluded, now on epistemological grounds, the possibility to interpret the Correspondence Principle as asserting the inclusion of classical mechanics within quantum theory.[53]

Jammer continues:

Contrary to Planck and Einstein, Bohr did not try to bridge the gap between classical and quantum, but, from the very beginning of his work searched for a scheme of quantum conceptions which would form a system just as coherent, on the one side of the abyss, as that of classical notions on the other. [54]

G. Tagliaferri also emphasises this point.[55] It can be argued that Bohr interpreted Cr as a merely formal mapping of CP onto QP. Let us call it Cr(2). On the other hand, as we know, this agreement founded on a mere mapping of one theory onto the other, cannot be extended to the whole of quantum physics. This would invoke Bohr's "fundamental contradictions" between classical and quantum theory.

CHAPTER 13

FROM THE 1926 WAVE MECHANICS TO A SECOND-QUANTISATION THEORY: SCHRÖDINGER'S NEW INTERPRETATION OF WAVE MECHANICS AND MICROPHYSICS IN THE 1950'S

13.1. Difficulties with the 1926 Wave Mechanics

Erwin Schrödinger (1887-1961) based his initial scientific program on a matter-waves interpretation of the psi-waves of his 1926 wave mechanics.[1] In March 1926 he proposed the so-called electrodynamic interpretation[2] (psi-square taken as proportional to charge density) and shortly afterwards, the beat-interpretation of atomic radiation emission[3] (atomic radiation as a beat phenomenon between stationary psi-waves, the emission-frequency equal to the beat frequency).

It can be assumed that the following new aspects of quantum mechanics (QM) influenced Schrödinger's decision to somewhat modify his former views: Max Born's statistical interpretation[4] of psi-squared appeared in the middle of 1926 and, to Schrödinger's disappointment, met with almost universal approval among physicists. Again in 1926, Schrödinger proved that his wave-mechanics and the Heisenberg-Born-Jordan matrix theory were mathematically equivalent.[5] Heisenberg's indeterminacy relations (IR) and Bohr's complementarity appeared almost at the same time, in 1927.

Schrödinger was highly impressed by IR. His decision to abandon the electrodynamic interpretation made public in a paper[6] in 1928 was justified, in fact, by the impossibility, after Heisenberg's relations, of a space-time description of micro-processes because these relations "have altered our conception of a world-image and what we should intend as its description".[7]

Another turning point was Schrödinger's 1931 interpretation[8] of the indistinguishability of micro-objects in the Bose-Einstein new statistics. Unlike his 1926 wave interpretation, he now thought that the lack of individuality in the atomic world was an aspect of a more general crisis in the ontology of classical atomism, a crisis also revealed by Heisenberg's relations. In fact, the new statistics, and the related indistinguishibility, represented for him the manifestation of an inner failure of the classical world-view and of its related objectification, and created a major incentive for a radical innovation in the conception of physical systems and of their states. As we shall see, he again took up these themes in his lectures and methodological papers in the fifties.

In 1935, Schrödinger published his famous essay,[9] better known as Schrödinger's Cat, in which Bohr's and Heisenberg's views on QM (the so-called Copenhagen philosophy) were challenged by showing that they led to an inconclusive paradox. His ideas were, on this last point, substantially in agreement with Einstein's EPR. In the same essay, however, Schrödinger criticised the statistical interpretation of QM in Einstein's sense. This little known detail[10] is one of the first instances in which Schrödinger differentiates his views from Einstein's position.

I consider the above developments in QM in the pre-war years as the background that came to affect, in the fifties, Schrödinger's novel outlook on physics and philosophy.

This outlook can also be traced to Schrödinger's early contributions to gas theory and wave mechanics, and to his epistemological interests in the thirties. Alexander Rüger has shown that Schrödinger's involvement with Eddington's program in the late thirties, and with unified field theories and QM in the thirties and forties, represented the background for his research on affine theory in the forties and for his new conception of micro-physics in the fifties.[11]

13.2. Schrödinger's 1950's Remedy for the Multidimensionality of the Psi Wave: Second Quantisation

In 1952, Schrödinger abandoned field theory with a pessimistic outlook towards its future; he had just begun a series of lectures and publications some of them at a non-professional level on epistemology and philosophy, collected and published under the title: *Science and Humanism* (1951). In 1954 he published *Nature and the Greeks*, in 1956 *What is Life? and Other Scientific Lectures,* followed in 1957 by *Science, Theory and Man,* and, in 1958, by *Mind and Matter.* Later he expressed his mature ideas on the foundations of modern physics, with special attention to the problems posed to quantum mechanics (QM) by the interpretation of the Copenhagen school.

What he proposed was a renewed view of wave mechanics, in which his early theories on the physical meaning of the psi-function and of the new statistics were now re-examined in the light of his new conception of physical systems. In this sense only, his views of the 1950's may be considered a return under a modified aspect to his conceptions of wave mechanics of the 1920's.

In the 1950's, Schrödinger's rejection of classical ontology and his new conception of physical systems found decisive support in the so-called *second-quantisation method* [12] of QM.

Since the early years of his wave mechanics, Schrödinger

was faced with the problem of the multidimensionality of the wave equation for a multiparticle system. In his *La Mecanique des Ondes* (1928), the multidimensionality problem is posed thus:

[Concerning that theory] which takes as its basis a wave-like phenomenon in the coordinates' situational space (q space)...it is clear that this space should be regarded as a mathematical artifice, an artifice which has also been used in ordinary mechanics. *In the end one describes even here eventsin physical space and time,* but, in effect, one has not yet succeeded in perfectly reconciling the two viewpoints [my translation from the French; my italics].[13]

It is clear that, for Schrödinger, physical space was at that time limited to three-dimensional space and, consequently, he was reluctant to consider the n-dimensional co-ordinates of situational space in his equations as a part of physical space. He found that approaches very similar to his own, i.e., using pluridimensionality as a mathematical artifice to be eliminated in the final formulation of his laws, "were found in Bohr's theory, when one builds up the theory of upper atoms starting from the hydrogen atom's trajectories".[14]

Following the fact that the introduction of a four-dimensional non-Euclidean space in Einstein's GR had produced such outstanding results, results well known to Schrödinger, it should have seemed to him obsolete to consider it as a mere mathematical artifice. Nonetheless, the idea that perceptual space is three-dimensional, and that a physical space cannot be other than the perceptual space, seemed rooted at that period in Schrödinger's thought.

In the same period, the fact that Schrödinger found it difficult to introduce imaginary quantities in his wave equation was an expression of the same attitude: in fact he tried by any means to avoid the appearance, in his equations, of the imaginary number "i".[15]

I consider these ideas of the twenties as symptoms of

Schrödinger's incomplete detachment from a so-called classical conception of physical theory, in spite of his rather innovative advances in other parts of physics at that time. Only later did Schrödinger understand the fundamental role of complex quantities in his wave equation and in the second quantisation method, as I will shortly show. The new understanding gave a major impetus to Schrödinger's theoretical work in the Fifties: he introduced a significant modification in his above-mentioned position by returning to a wave theory of micro-physics which included both Heisenberg's IR and the new statistics.

In his 1950 article, "What is an elementary particle?" Schrödinger examined in some detail the various interpretations of Heisenberg's IR, finding them unsatisfactory because, until then, no picture had been found of anything that complied with the IR requirements on conjugated variables. He believed that this lack of picturability could not relieve the physicist from a search for "a real understanding". He then found a way out of the difficulty: for the Copenhagenists, IR referred to the particle, but the particle was not an identifiable individual entity:

> It may indeed well be that no individual entity can be conceived which would answer the requirements for the adequate picture stated above.[16]

He added that the vagueness of Born's probabilistic interpretation concerning the question whether the wave gives information about one particle or about a collection of particles, is to be taken as a symptom of the difficulty inherent in keeping intact the particle concept and, at the same time, conceding that the particle possesses no individuality (i.e., the concept of an indistinguishable particle):

> It is not easy to realise this lack of individuality and to find words for it. The

> probability interpretation ...seems to be vague as to whether the wave gives information about one particle or about an *ensemble* of particles.[17]

He continued by stating that there existed exact mathematical tools for distinguishing between the two alternatives above, i.e., finding out which is the referent of the information given by the wave. One of these tools was indicated by him in 1926, but, in 1950, he superseded his method with a better approach:

> A method of dealing with the problem of many particles was indicated in 1926 by the present writer....Deeper insight led to its improvement... the many-dimensional treatment has been superseded by the so-called second quantisation, which is mathematically equivalent to uniting into one three-dimensional formulation the cases N= 0,1,2,3....(to infinity) of the many-dimensional treatment. This highly ingenious device includes the so-called new statistics....It is the only precise formulation of the views now held, and the one that is always used. What is very significant in our present context is that one cannot avoid leaving indeterminate the number of the particles dealt with. It is thus obvious that they are not individuals.[18]

Notice that one reason for Schrödinger's acceptance of second quantisation is to be found in the new statistics which are, as it is known, a remarkable by-product of this quantisation.

Again in 1953, in a paper written on the occasion of the celebration of De Broglie's sixtieth birthday, held at the Henry Poincarè Institute in Paris, Schrödinger confronted the problems of indistinguishability and multidimensionality. He began his paper by remarking that he himself and De Broglie were shocked and disappointed when "[they] learnt that a sort of transcendental, almost psychical interpretation of the wave phenomenon has been put forward".[19]

Against this interpretation, which was evidently the Copenhagen interpretation of QM, Schrödinger counterpoised his and De Broglie's own interpretation via the wave-function. In response to the accusation that his wave-like interpretation "seemed deceptive and, after all, too naive", Schrödinger agreed to reconsider it in the light of "two new aspects which have since arisen".

He thought that these aspects were "very relevant for reconsidering the interpretation". He listed them as: a) indistinguishibility of particles, following the new statistics; b) second quantisation procedures. He added that "they [the two aspects] are intimately connected...[and] have not turned up suddenly. Their roots lie far back, but their bearing was only very gradually recognised".[20]

In his understanding, the indistinguishability aspect a) leads to the non-individual identifiability of "what has been called a particle" and, consequently, to the rejection of the corresponding concept. Renouncing the particle concept eliminated the oft-mentioned difficulty of the spreading out of the wave-packets.

In the same passage he quoted papers of his own where "he dwelt at length on this point".[21]

Concerning b), he stated:

If a particle is not a permanent entity.....quantization of the De Broglie Waves around a nucleus welds into one comprehensive scheme all the 3n-dimensional representations that I had proposed for the n-body problem.[22]

Again, he considered superseded his "naive" 1926 approach:

[The lamented inconsistency with his wave theory] will be avoided by returning to a wave theory that is not continually abrogated by dice-miracles; not of course to the naive wave theory of yore, but to a more sophisticated one, based on second quantization and the non-individuality of particles.[23]

It is remarkable that since 1950 Schrödinger recognised that his 1926 approach to the many-particle problem was superseded by second quantization, an approach, after all, which partially followed Heisenberg's matrix-method, which he disliked.

13.3. Schrödinger's Original Conception of the New Statistics

Early in his scientific life, Schrödinger emphasised that the interpretations of the Bose-Einstein statistics in the sense the indistinguishability of particles, was for him an indication that it is the classical particle concept which needed revision. This conclusion strengthened Schrödinger's belief that the classical particle concept had no more right to exist and that second quantisation was the most apt mathematical tool for his theory (a mathematical technique, incidentally, which was developed by his opponents, the damned Copenhagenists).

Schrödinger dwelt at length on this matter in his popular article, in order to illustrate the reversed roles that in his new theory were attributed to the concepts of particle and to energy state. On such occasions, he was always please to present very attractive metaphors and similes (Appendix 1, of this chapter).[24]

One can argue that Schrödinger accepted second quantisation because the technique confirmed that the referents of the new statistics were not time-honoured classical particles, deprived of just some of their particulate properties. His second-quantisation approach denied the old conception that a macroscopic body resulted from an assemblage of classical particles, simply endowed with the new-statistical property of indistinguishability.

It is remarkable that second quantisation also provided a key for solving Schrödinger's well-known paradox of his cat being simultaneously dead and alive. The cat and of the cat-killing machine cannot be simply reduced to the wave-like properties emerging from the first-quantisation process (or from the mathematically equivalent properties of matrices).[25]

Second quantisation was thus a remedy not only for the difficulties of wave-packet spread and wave multidimensionality in the first-quantised theory, but also helped to eliminate the paradoxical aspects of the cat's story (Appendix 2, of this chapter).

His original approach to the new-statistics goes back to his early years: in his 1922 paper, "Was ist ein Naturgesetz?",[26] his *Antrittsrede* to Zürich University, he remarked that, although the conformity to the statistical laws of the average values of gas properties is usually considered as the necessary consequence of the random elastic collisions of the particles of the "ensemble" (following casual mechanical laws), this necessity can, in effect, be disproved. For it would suffice to admit that the particle energy-state could fluctuate with equal probability by a positive or negative amount.

The kinetic theory hypothesis and the related theory of fluctuations is there considered a sufficient but not a necessary condition for a statistical-type gas theory. In an apparently naive but, in effect, ironical mood, Schrödinger asks why the gas pressure is not reduced to the sum of the pressures of the single molecules. If the answer is that this would be too naive a reductionism, then, Schrödinger seems to imply, mechanistic reductionism of the gas fluctuation does not fare much better.

The last remark represents a subtle but radical criticism of the well established interpretation of classical statistics and of its foundations in atomism and mechanistic reductionism.[27]

Schrödinger's arguments above implied that the same criticism would be appropriate to the atomism and the old-fashioned determinism of the Copenhagenists. In his essay "Über Indeterminismus in der Physik" (1931) published in his 1932 book of the same title,[28] he remarked that his teacher Exner had superseded determinism as early as 1918, nine years before Heisenberg proposed his IR.

At bottom Schrödinger refused in these claims to accept the idea of a statistical ensemble as the necessary and sufficient basic conception for a theory of the statistical character of measurement. On this point he was consistent with his ideas of the thirties.[29]

13.4. Schrödinger's Criticism of Atomistic Ontology in 1950

Schrödinger's four public lectures, *Science as a Constituent of Humanism*, were delivered under the auspices of the Dublin Institute for Advanced Studies in February 1950 and published under the title *Science and Humanism*.[30] I consider this booklet to be a reliable summary and a consistent development of themes which can be traced back to Schrödinger's early contributions to gas theory and wave mechanics, to his epistemological interests in the thirties, and to his contributions to *affine theory* in the forties.

In this respect, Schrödinger's ideas in *Science and Humanism* set out the foundational conceptions for his technical approach through second quantisation. In his lectures he criticised the commonly accepted view that atomism is a condition imposed upon theory by facts and experiments; in his own words "discontinuous aspects of nature (are not) forced upon us very much *against our will*" (italics in original).[31] In his view, Atomic theory is not the simple result of "our refined experiments", but it stems from an effort to explain very basic experiences like the variation of volume in spite of the conservation of weight and mass. This basic feature also explains why ancient Greek philosophers could guess the atomic constitution of matter even if they lacked the much more refined experimental techniques of our times.[32]

In contrast with the accepted view (held even by the majority of scientists) "that their [i.e. the Greeks] atomic theory has been a completely unfounded guess which might just as well have turned out to be a mistake",[33] Schrödinger maintains that ancient atomism was, theoretically and experimentally, a well-founded theory. It represented a theoretical explanation of the very basic experience of material objects changing shape (in the processes of fusion and gasification), while conserving their weight, because only a body composed of atoms and void can expand or contract without creating or destroying its matter.[34] In supporting this view by historical references, Schrödinger shows his deep interests in the historical perspective of classical science, the theme of another of his booklets, *Science and the Greeks* :

Our notion of the elementary particle has historically descended from their [i.e. from the ancient Greek Philosophers'] notion of the atom and is conceptually derived from their notion of the atom: *we have simply held on to it* [italics in original]. [35]

However, despite its apparently factual foundations, atomism is in essence an idea, not a fact; albeit an idea with a profound history. In fact, the choice of the atomic paradigm was dramatic for the Greeks because, as Schrödinger illustrates in other parts of his work, Greek mathematical philosophy also had, in contrast with the discontinuous features of atomism, a deep concern with continuity. Identity means continuous conservation of shape (not just of substance), and it also represents for Schrödinger one of the pieces of evidence supporting our conception of nature.[36]

Continuous conservation of shape, however, presented, in Schrödinger's view, conceptual difficulties, and this explains the endurance of the atomistic paradigm:

It would thus seem that physical science in its present form in which it is the direct offspring, the uninterrupted continuation, of ancient science was from its very beginning ushered in by the desire to avoid the haziness inherent in the conception of the continuum, the precarious side of which was then more felt than in modern-times, until quite recently... *this explains why atomism has proved so successful and durable and indispensable* [italics in original].[37]

13. 5. A Critique of the Copenhagen Philosophy

Having dismissed the thesis that atomism is imposed by facts, Schrödinger is now in a better position to discuss its present difficulties. He emphasises the point that the main feature of classical atomism, distinguishability (DT), is denied by the new statistics. If the concept that atoms can be distinguished and counted is no longer tenable, his argument runs, atomism has lost one of its main conceptual planks:

It seems almost ludicrous that precisely the same years or decades which let us succeed in tracing single, individual atoms and particles (and that in various ways), we have yet been compelled to dismiss the idea that such a particle is an individual entity which in principle retains its sameness for ever.... There are cases where the sameness becomes entirely meaningless.... I beg to emphasise this and I beg you to believe it. It is not a question of our being able to ascertain the identity in some instances and not being able to do so in others. It is beyond doubt that the question of sameness, of identity, really and truly has no meaning....Atoms our modern atoms, the ultimate particles must no longer be regarded as identifiable individuals. This is a stronger deviation from the original idea of an atom than anybody had ever contemplated....[38]

According to Schrödinger, physicists have wrongly assumed that distinguishability of atoms was a property of natural objects and not a model of our theorising.[39] They believe that conservation of substance is imposed by facts, while in effect it is really an assumption (a habit of language) and, consequently, it makes no logical sense to transfer it from macroscopic to microscopic models. Conservation of form (or of shape; German: *Gestalt*) is for Schrödinger a more basic experience (i.e., more consistently founded on the perceptive level) than substance conservation, which seems uncritically bound to the assumption of a distinguishability of atoms:

> The habit of everyday language deceives us and seems to require, whenever we hear the word "shape" or "form" pronounced, that it must be the shape or form of something, that a material substratum is required to take on a shape. Scientifically this habit goes back to Aristotle, his *causa materialis* and *causa formalis*. But when you come to the ultimate particles constituting matter, there seems to be no point in thinking of them again as consisting of some material. They are, as it were, pure shape, nothing but shape; what turns up again in successive observations is the shape, not an individual speck of material. The new idea is that what is permanent in these ultimate particles or small aggregates is their shape [German: *Gestalt*] and organisation.[40]

In short, since atomism and continuity have long been fundamental (although controversial) perspectives of science, classical physics should not to be understood as the natural home of contiguous action, and discontinuity is not the scandal of QM. The Copenhagenist views of Heisenberg and Bohr who presumed that discontinuity, was a conception irremediably cast upon us by modern experiments, are thus challenged by Schrödinger on historical and philosophical grounds. Besides, as we shall see, his critique concerns specific aspects of the Copenhagen philosophy.

Schrödinger's criticism of the Copenhagenists' supposed solution to the problem of the interaction between the observer and the observed system, or between subject and object, follows the same philosophical and historical approach of his former arguments. He starts by discussing a slightly different and obviously connected problem: is the new physical theory of quantum mechanics "*deeply rooted in the facts of observation....* or is this new aspect (of theory) perhaps the mark, not of objective nature, but of the setting of the human mind, of the stage that our understanding of nature has reached at present?" [my italics].[41]

The answer to this question is extremely difficult because "it is not even absolutely clear what this antithesis means: objective nature and human mind".[42] Schrödinger refers to the problem of the subject-object relation as it has been tackled by the philosophical tradition which culminated in Kant, and he realises the difficulties of a definite answer.[43]

Schrödinger questions again the inductivist's axiom by asking the question: "Is the impossibility of a continuous, gapless, uninterrupted description in space and time really founded in incontrovertible facts?" He answers in the negative. But he also acknowledges that Bohr and Heisenberg meant something different with their form of inductivism. When stating that the cognitive subject and the object are correlated in the process of acquisition of knowledge, they meant that the very act of observation changes the nature of the object, and, presumably, the effectiveness of the theoretical prediction. Theory then becomes useless for a second prediction. According to the Copenhagenists, theory and observations are thus interdependent in a sense different and more radical than the one provided by the inductivist's philosophy.

Schrödinger's objection to the true Copenhagen philosophy is itself very radical. His argument transcends the strict (epistemological) level to which Bohr's and Heisenberg's considerations were confined. He derives his inspiration from broad philosophical and cultural perspectives:

I cannot believe that the deep philosophical enquiry into the relation between subject and object and into the true meaning of the distinction between them, depends on the quantitative results of physical and chemical measurements with weighing scales, spectroscopes, microscopes, telescopes, with Geiger-Mueller-counters, Wilson-Chambers, photographic plates...and what-not. It is not easy to say why I do not believe it. I feel a certain incongruity between the applied means and the problem to be solved. I do not feel quite so diffident with regard to other sciences, in particular biology, and quite especially genetics and the facts about evolution....[44]

Schrödinger's objections to the Copenhagen conception of the interaction between the observer and the observed system receives further support from his foundational assumption that: *"the observing mind is not a physical system, it cannot interact with any physical system"*[45] [italics in original].

The incongruity that he believes exists between the subject's consciousness and the world of his scientific knowledge is reinforced by his objection to a well-known attempt (made originally by Jordan) to explain freedom in moral conduct by extrapolating to the latter the QM indeterminacy of physical laws. Schrödinger's objection is drawn from the philosophy of Cassirer, who held that the two antagonistic concepts (in physics!) of freedom and necessity lose their explanatory power in matters of moral conduct, since moral behaviour, as Schiller has splendidly illustrated in his *Wallestein*, is both compulsory and free.[46]

In short, Schrödinger maintains that philosophical problems cannot be resolved by a physicalist approach. This is consistent with his philosophy, because elsewhere in the same work he states that, beyond physics, other fields such as art and literature help expand human knowledge.[47]

13. 6. Continuity and Causality not at the Same Foundational Level: a Two-Level Theory

Though arguing for this radical change in the ontology of classical physics, Schrödinger thought he had remained within the rationalistic tradition of physics and of western thought in general, quite different from the strange mixture of psychical and physical features[48] that the Copenhagen interpretation of the psi-collapse demanded.

"Continuity of description" was, according to him, the founding postulate of classical physics, not to be abandoned in the new conception of theory.[49] Schrödinger thus introduced the principle of continuity and the causality postulate:

> The exact physical situation at any point P at a given moment t is unambiguously determined by the exact physical situation within a certain surrounding of P at any previous time, say t-dt. If dt is large, that is, if that previous time lies far back, it may be necessary to know the previous situation for a wide domain around P. But the "domain of influence" becomes smaller and smaller as dt becomes smaller, and becomes infinitesimal as dt becomes infinitesimal. Or in plain, though less precise, words: what happens anywhere at a given moment depends only and unambiguously on what has been going on in the immediate neighbourhood "just a moment earlier".[50]

Schrödinger's continuity principle excluded causal connection through instantaneous action at-a-distance and claimed a finite propagation velocity for this connection, although it did not prescribe precise relativistic limits for the velocity. It claimed that a causal connection had a non-ambiguous meaning only at infinitesimal space-time distances (local causality). The possibility of conceiving infinitesimal space-time distances, the continuity requirement, is thus a prerequisite for a non-ambiguous definition of causality. This means that the postulate of continuity (COP) was considered as a necessary prerequisite but not a sufficient condition for causal connection (CAC):

> Obviously, if the ideal of continuous, "gapless", description breaks down, this precise formulation of the principle of causality breaks down.[51]

COP was a pre-requisite for theory on a meta-theoretical level: a theory which claims to be precise, logical and clear (i.e., complete in Schrödinger's sense) cannot renounce COP because thinking (theorising) has exigencies of its own:

> From an incomplete description from a picture with gaps in space and time one cannot draw clear and unambiguous conclusions; it leads to hazy, arbitrary, unclear thinking.[52]

Thus, COP has a more fundamental status than the CAC postulate because the former represents the mental frame through which we try rationally to shape the interrelations of phenomena. The ideal continuous description, a continuity in thought, is essentially different from the imperfect continuity that one experiences on the perceptive-descriptive level; the latter, in fact, is mere approximation as, for example, when we measure quantities with the maximum of accuracy in a short interval.[53]

Among the foundational ideas which support second quantization, special attention must be paid to Schrödinger's mature view of the ways in which concepts are linked with the observable facts, a problem in TP we have already encountered.

His conception of physical theory requires, first of all, that logical coherence and completeness of language on the theoretical level are contrasted with the incompleteness and causal "gaps" of language at the descriptive observational level. Consequently, a (purely) theoretical level conforming to COP cannot claim to be a description of observables:

> We do give a complete description, continuous in space and time, without leaving any gaps, conforming to the classical ideal of a description of something. *But we do not claim that this something is the observed or observable facts* [my italics].[54]

It follows that a bivocal correspondence between the terms of the two languages is not possible:

> The gaps eliminated from the wave picture have withdrawn to the connection between the wave picture and the observable facts. The latter are not in one to one correspondence with the former.[55]

The postulate of causality (CAP) must keep its validity at the theoretical level in spite of the fact that it may break down at the level of observation:

> We must not be astonished to meet in this order of ideas with new, unprecedented difficulties as regards causation. We even meet (as you know) with the statement that there are gaps or flows in strict causation.[56]

As an example of such a situation Schrödinger mentions "Bohr's famous theory of spectral lines in 1913, [which] had to assume that the atom makes a sudden transition from one state into another.... No information about the atom during this transition can be offered".[57] This means that the transition itself is not observable in principle; there is a gap in the observational language. This lack of information, a gap in continuity on the descriptive level (an "incompleteness in description" of the observational language) does not, however, prevent there being a continuous purely theoretical language.

Since, on the level of a pure, complete and continuous theory, language does not refer to observations, theory presents itself in the true model that we approach gradually, without perhaps ever reaching it, owing to human imperfection".[67]

The real world, as such, is unthinkable (almost as a triangular-circle), but this does not imply that the same real world is not practically accessible (the observatonal level has an autonomy of its own). Fruitfulness in research is achieved if the scientist correctly intends the pursuit of a clear model to be a search for a *Bild*. In this case, the achievement of clarity will coincide with that of adequacy.

On these matters Schrödinger believed he kept faith with Boltzmann's ideas:[68]

Ludwig Boltzmann strongly emphasised this point; let me be quite precise, he would say, childishly precise about my model, even though I know that I cannot guess from the ever incomplete circumstantial evidence of experiments what nature really is like. But *without an absolutely precise model thinking itself becomes imprecise, and the consequences to be derived from the model become ambiguous* [my italics].[69]

On the observable level one may remain pragmatic:[70] the successes of QM essentially prove this. But pragmatic considerations become scientific only when guided and illuminated by an adequate philosophy. This is feasible in as much as theories are *Bilder,*[71] conceptual models whose main scope is to "confirm expectations".[72] Therefore their validity then should not be judged by their truth-value, but by their adequacy:

We plan...the pictures...for the purpose of seeing whether they confirm expectations thus whether the expectations were reasonable, and thus whether the pictures or models we use are adequate. Notice that we prefer to say adequate, not true. *For in order that a description be capable of being true, it must be capable of being compared directly with actual facts. That is usually not the case with our models* [my Italics].[73]

In one of his last lectures in Vienna in 1958, published in

form of a conceptual model (*Bild*) and language now has a merely interpretative role.

A distinction, more precisely an opposition, is thus highlighted between the observational descriptive and the theoretical levels: "the observed facts are repugnant to the classical ideal of a continuous description in space and time".[58]

The view of a two-level conception was not convincing for Heisenberg, or, perhaps he never seriously tried to understand Schrödinger's new conceptual framework.[59]

13.7. Only the Ondulatory Theory is a "Complete" Theory

An illustration of the views above is Schrödinger's interpretation of the experiment on the diffraction of the electron, a well known case in QM, as it was the object of a long controversy between Einstein and Bohr. Since the concept of an electron as a particle whose motion can be described in space-time needs to be revised in consequence of the diffraction pattern and of Heisenberg's IR, Schrödinger suggested that the unquestioned validity of the ondulatory language should be fully exploited in this case. He justified his proposal with the remark that "the concept of waves is unavoidable" both in the diffraction experiment with cathode rays (i.e., electrons) and in the diffraction experiments with photons (i.e., electromagnetic waves). One could in turn observe that the reverse is also true, i.e., that the particulate aspect seems "unavoidable" with either one or the other of the two experiments (*pace* Bohr). Schrödinger would reply by remarking that, due to his interpretation of the new statistics as representing a crisis in distinguishability and to the role of COP, his new wave-theory is more adequate than a theory of particles.

In a special instance, he insists that the results of electron diffraction and other similar experiments are fatal to the particulate aspect. For in the two-slits diffraction experiment one does not find the correct particle density distribution that would be expected as a consequence of the particle superposition principle, but one does find the correct superposition for the wave-amplitude intensity:

> The wave picture conforms with the classical demand for complete determinism, the mathematical method used is that of field-equations, though sometimes they are a highly generalised type of field-equations.[60]

To the objection that a stable connection between his wave function and the space-time description for the behaviour of the correlated particle has not been found, Schrödinger would answer that any such theory is not necessary because the absence of a connection "between the wave picture and the observable facts" is not an obstacle. A complete theory is required only for the waves (in accordance with the COP), while an incomplete connection between the theory and the causal behaviour of the particles is acceptable. Clearly, Schrödinger here restates his view of a discrepancy between the two levels of a pure theory and the language of observables.

In one of his 1950 essays, "What is an elementary particle?", Schrödinger discussed the significance of the IR, and criticised the Copenhagen view because it is based on the ontology of the particle. In contradiction with his underlying premise, it affirms, however, that IR forbids complete information on the magnitudes of the particle:

> We must not believe that the more complete description they demanded about what is really going on in the physical world is conceivable, but in practice unobtainable....*We have taken over from previous theory the idea of a particle and all the technical language concerning it. This idea is inadequate*. It constantly drives our

mind to ask for information which has obviously no significance....The uncertainty relation refers to the particle. The particle, as we shall see, is not an identifiable individual [my italics].[61]

The practical impossibility of a simultaneous measurement of conjugated quantities (following IR) and its interpretation as a lack of information about the object, was contrasted by Schrödinger with the potentiality "of an idea of the physical object formed in our mind (*Vorstellung*), that contains in some way *everything that could be observed in some way or other by any observer, and* not only the records of what has been observed in a particular case".[62] [my italics].

This potentiality *in principle* was related to that "completion of facts in thought" (*Ergänzung der Tatsachen in Gedanken*), a view Schrödinger attributed to Mach and one which he exemplified by referring to "our grasp of a shape of a solid by visualising it in a three dimensional space...even though the eye can at any moment only perceive one perspective view".[63]

Due to the exigency of completion, "the question what is now the wave function (meaning, what is now the actual state of the physical system?) must be regarded as meaningful, even though it can hardly ever be answered exhaustively".

Clearly, Schrödinger's adherence to the ondulatory conception represents a theoretical choice closely related to his epistemology: the conception of a pure continuous theory. This, the most important feature of Schrödinger's mature thought in the 1950's, and should be considered as Schrödinger's intellectual testament.

13.8. A Purely Theoretical Language Needs a Clear and Precise Model

In Schrödinger's above conception of a pure theoretical language, only the adoption of absolutely clear and precise models can satisfy the condition for an adequate theory. In order to understand the meaning of a clear model, he points out that a physicist might be misguided in the pursuit of "a clear picture" should he mistake such a pursuit for a search of a "true theory", i.e., a true description of nature.[64] In order to have the advantage of a clear model, there is a price to be paid in the renunciation of a real description of nature. Let us consider Schrödinger's specific issue :

> But we do not claim that this something [i.e., the model] is the observed or observable facts; and still less do we claim that we thus describe what nature (matter, radiation, etc.) really is. In fact we use this picture (the so-called wave picture) in full knowledge that it is neither.[65]

The misinterpretation of a macrosopic model as a true theory would be fatal to the theory's success :

> No model shaped after our large-scale experience can ever be true...We find nature behaving so entirely differently from what we observe in the visible and palpable bodies of our surroundings.... A complete satisfactory model of this type is not only practically inaccessible, but not even thinkable. Or, to be precise, we can, of course, think it, but however we think it, it is wrong; not perhaps quite as meaningless as a "triangular circle", but much more so than a "winged lion ".[66]

From the view that no model can be completely adequate, we are not authorised to conclude that models are useless: it is only through models that we can think theoretically. Later Schrödinger specified that the "absolutely precise [model] could be misinterpreted as the true model...[which] exists in the Platonic realm of ideas a

Nuovo Cimento, [74] Schrödinger returned to some of the ideas above; against "prevailing opinion", that the Compton effect represented a demonstration of energy and momentum conservation in QM (a proof, as such, of the particulate conception), he reinstated his 1927 theory,[75] where he had shown that the Compton effect can also be explained by a wave-like theory (the same problem he had treated in his Dublin Lectures).

A yearning for a complete description of the material world in space and time was again expressed in Schrödinger's last essay discussed above, and he still considered it far from proven that this ambition cannot be reached.

13.9. Schrödinger's New Ideas in the 1950's: a Programme for New Physics

I think that an overall judgement on the nature and scope of Schrödinger's scientific work in the fifties could be encapsulated by noting that: he wanted to keep open the possibilities for a future continuous theory of microphysics which could oppose the then triumphant Copenhagen philosophy. In this sense, his work in the fifties represented the fulfilment of what he considered his intellectual mission. As an initial approach, it was, for him, important to clarify the general epistemological context of this possible future theory and to develop a radical critique of the Copenhagen philosophy.

In an effort to explain his ideas to a more general audience, not just physicists, he often used similes and metaphorical illustrations. Consequently his papers were predominantly verbal in form, even when they dealt with abstract physical and mathematical concepts. He frequently referred to his or others' former works to provide the technical and mathematical support for his theories.

To the possible objection[76] that in presenting a radically new physical theory (as Schrödinger did in his above-mentioned writings) he could not return to his old 1926 conceptions of psi-waves, my answer is that Schrödinger's work, since his early years, manifested a remarkable variety of viewpoints and approaches,[77] not all in line with his 1926 ideas on psi-waves. Though I do not presume that my list is complete, I have shown in this essay how certain innovative themes of the 1950's were announced in some of his earlier writings.

Schrödinger was aware of the enormous effort and devotion required in order to produce significant innovations within the predominant paradigm of contemporary physics, an effort which can only rarely be the task of one man. In 1952, somehow confessing his impotence, he complained:

I beg to plead that I am at the moment groping for my way almost single handedly, as against a host of clever people doing their best along the recognised lines of thought.[78]

His intellectual isolation from the world centres of the scientific enterprise in the most fruitful period of his life was also a consequence of the events he experienced in the pre-war and war years, and of his opposition to Nazi power. This is an accepted fact. However, one should concede that his isolation began earlier, around 1926, when he began to oppose the dominant Copenhagen interpretation of QM. This leads to a number of other questions: was this isolation the result of those unquestionable answers which nature supposedly dispenses to its scientific adepts, or was it just "largely a sociological accident"?.[79] In the latter case, could the fact that contemporary physicists abandoned Schrödinger's theory and modern physics proceeded in the way it has be attributed to a sociological accident as well ?

Historians may find answers to such questions through a detailed study of the available sources and a search for the unknown ones. Physicists can also help to improve Schrödinger's historiography. In the last decade, Schrödinger's idea of a wave theory of microphysics returned to the forefront of some physicists' interests. Various attempts were made to revive Schrödinger's interpretation of QM, amending what were considered to be the defects of his former theory.[80]

However, to my present knowledge, no one followed the path mapped out by Schrödinger's papers in the 1950's. It can be assumed that, were he alive, Schrödinger would not have considered the majority of the amendments above to be consistent with his theoretic conception in physics.

These remarks might perhaps suggest to historians that they should pursue (or attempt to) a reconstruction of a wave theory of microphysics in accordance with Schrödinger's later views. In this case, a deeper analysis of his own and of other theories, as indicated in his papers, would become indispensable. Apart from the possibility of its acceptance today as a valid theory, this reconstruction would certainly contribute to an improvement in our understating of the physical and epistemological ideas of one of the most outstanding scientist-philosophers of our time.

Appendix 1. New Statistics and the Demand for a New Objectification: Schrödinger's Simile Concerning Three Schoolboys and their Three Awards.

In a popular mood, Schrödinger presented[81] his readers with a striking example of the fact that different statistical approaches are related to different (from the usual ones) ways of selecting physical systems (the objects) and their states. Schrödinger explains this with the following simile: three schoolboys deserve rewards and the available awards are of three different kinds:

Two portraits, one of Newton and one of Shakespeare.
Two shilling-pieces, each piece an indivisible quantity.
Two vacancies on the school football team.

According to the kinds of awards, different distributions of the same among the boys are possible:

Awards	Distinguishability of awards	Cumulativeness	Number of distributions	Statistics
Two portraits	Yes	No	9	Boltzmann
Two shillings	No	Yes	6	Böse-Einst.
Two vacancies	No	No	3	Fermi-Dirac

In Schrödinger's simile, the common-sense objectification is somehow reversed because boys represent states of the system and the allocation of awards represents objects. In fact, a football team vacancy offered to one boy named Dick means: the particle "vacancy" takes on the "state Dick".

The identity of the three boys is thus somehow restricted, reproducing in the simile Schrödinger's conception of a restricted notion of physical objects' identities. Notice that the noncumulativeness of awards, in cases 1 and 3, illustrates, in the simile, Pauli's exclusion principle.

Another of Schrödinger similes conveys the following metaphorical meaning: if, according to classical physics, one wishes to represent positions in a bank as corresponding to the energy-states of particles and the employees themselves as classical particles, then Schrödinger's new conception compels one to reverse the correspondence, i.e., to liken the employees' positions to the new-statistical objects and the staff members themselves to energy states.

Appendix 2. Schrödinger's "Fuzzy" Cat is not Representable by a Cloud but by a Factorisable Wave Functional

Schrödinger's equation for the case of N identical particles is:

$$i\frac{h}{2\pi}\frac{\partial \psi(1,2,...N;\ t)}{\partial t} = H(1,2,...N)\ \psi(1,2,...N;\ t)\ .$$

The solution for psi in the equation above is not factorisable. This fact expresses mathematically the physical situation of the actual "fuzzy behaviour" of the Schrödinger cat's psi in a the first-quantisation theory. According to Schrödinger, the cat's description given by the psi function is fuzzy (blurred, indistinct) for the reason that it is not possible to represent the cat's state as the superposition of an ensemble of cats' (p, q) states, with the cat decisively either dead or alive. This last would be Einstein's statistical interpretation of the cat's tragic destiny. According to Schrödinger, the fact that the cat's psi does not factorise in the first-quantisation theory (i.e., when its independent variables are the conjugated quantities p, q) is the mathematical counterpart of the impossibility of considering the cats' state a superposition of live and dead cats.

The fact that the psi is "fuzzy" does not necessarily imply that the cat consists of an ensemble of cats:

This prevents us from continuing naively to give credence to a "fuzzy model" as a picture of reality. There is a difference between a blurred or out of focus picture and a photograph of clouds and patches of fog.[82]

Schrödinger meant that a fuzzy snapshot is not always the superposition of clearly focused snapshots. Let us remember that Schrödinger made a similar remark about the interpretation of a gas state in kinetic theory (see above). For Schrödinger, the non-factorisability of psi is the consequence of a wrong representation: physical situations concerning the actual statistical behaviour of elementary particles cannot be reproduced by any mathematical theory that represents them as identifiable things. A restricted notion of identity is imposed by Schrödinger's view of QM.

The most apt mathematical theory for such state of affairs is *second-quantisation*, .i.e., a mathematical formalism[83] in which *the occupation numbers $N^1, N^2 ... N^i$ of the states* (not the co-ordinates of the particles, as in Schrödinger's 1926 equation) play the role of independent variables. A wave functional Ψ factorises under the above conditions. In fact, in the case of Böse particles, the expression for Ψ is:

$$\Psi_{N_1 N_2 ..... N} = (\sqrt{N_1! N_2! ... /N!}) \Sigma \phi_{p1}(\chi_1) \phi_{p2}(\chi_2) ... \phi_{p_n}(\chi_n) ,$$

N_i is the number of particles in the state ϕ_i. N_i may, of course, be zero. Here $p^1, p^2, ...p^n$ are the ordinal numbers of the states in which the individual particles are, and the sum is taken over all permutations of those suffixes $p^1, p^2, ...p^N$, which are different.

In my chapter on Schrödinger, I have shown that second-quantisation is the theory Schrödinger proposed in the 1950's.

CHAPTER 14

CONCLUSIONS

Throughout the second half of the nineteenth century in Europe, physicists such as Helmholtz, Maxwell, Hertz, Mach, Boltzmann, Planck, Poincaré' et al. were all aware of the philosophical problems implicitly present in the theories of physics. Through their *reflective criticism*, documented in a rich epistemological literature unequalled in our times, they made a major contribution to the process of shaping that peculiar mentality which distinguished the theoretician from the experimentalist. This trend continued into the first decades of our century.

By implicitly accepting the philosophical tradition of German physics, Einstein showed himself to be in favour of a necessary interpenetration of science and philosophy. He wrote in his "Reply to Criticism" :[1]

Epistemology without contact with science becomes an empty scheme. Science without epistemology is in so far as it is thinkable at all primitive and muddled.

Although he once labelled the scientist a philosophical opportunist, he meant by this apparently contradictory statement that a scientist does not always need to be committed to one of the established philosophical schools. Einstein was a true representative of a European tradition in which science and philosophy were often linked by the same aims.

Clearly, these statements contradict the ideas of those who consider the philosophical movement of the last two centuries as an epiphenomenon with respect to real science, i.e., something marginal in comparison with the great achievements of theoretical and experimental physics in this same period of time.

It is evident that the riddle of Cr for Einstein concerned the fundamental epistemological problem of identifying a criterion that distinguishes mathematical theories and theories of physics. And this should be no surprise if one argues, as I have, that the problem of a distinction/criterion between mathematics and physics theories was one of the central problems in Einstein's epistemology, as is clearly appearent in 1949 from the closing passages of his "Reply to Criticism" quoted above.

The fundamental role of Cr in modern physics is also confirmed by Bohr's work on related problems. We saw that the transition from the "restriction" to the "limitation" themes links Cr and Cm in Bohr's papers of 1925-1927. In 1927 he was concerned with the relation between definitions of theoretical concepts and observations. He found that an antimony exists between the conditions for definition and those for observation and that this antinomy imposes a new type of "restriction" on theories of physics, the "limitation" which is implicit in his Cm argument. The electron can be classically described either as a wave or as a particle (according to different experimental settings), but the two CT descriptions are complementary, i.e., they cannot be consistently adopted as part of a unique CT. In this sense Cm represents a "limitation" in the context of CT.

As we saw, this contradiction was also Schrödinger's concern in the fifties. He called "the classical ideal of uninterrupted continuous description" on both the observational and theoretical levels, the "old way",[2] meaning, of course, that this ideal was no longer attainable. He also acknowledged that the questions above were at the centre of scientific debate in the nineteenth- and twentieth- centuries:

> Very similar declarations...[were] made again and again by competent physicists a long time ago, all through the nineteenth century and the early days of our century...they were aware that the desire for having a clear picture necessarily led one to encumber it with unwarranted details.[3]

These competent physicists are almost certainly Hertz, Boltzmann and their followers.

One can thus argue that Schrödinger's two-level conception is, at bottom and despite its "amazing" appearance, part of the tradition of the nineteenth-century *Bild*-conception of physics,[4] formulated by Hertz in his 1894 *Prinzipien der Mechanik*, and also discussed by Boltzmann, Einstein et al. Schrödinger partially inherited the two-level conception from his teacher Exner, and he deepened his conception through his intense study of Boltzmann's work. In his oft-quoted dictum, Boltzmann asserted that "only one half of our experience is ever experience".[5]

Schrödinger's views concerning the status of theories brings to its culmination that tradition (to which Hertz and Boltzmann were major contributors) which values above all the predictive power of theories and plays down the problem of their truth-value as veridical descriptions of an objective reality (or even of phenomena); in this sense he is also against any form of phenomenalism.

In this tradition, interpretation, rather than description, was taken as the characteristic feature of theories. Consequently, adequacy, not truth, was the criterion for judging a theory's validity. The "partial congruence" of theory with "nature", a form of adequacy, which the *Bild*-conception supports, is consistent with Schrödinger's demand, in the 1950's, that a continuous field-type theory be compatible on the observational level with causal gaps and space-time discontinuities, expressed in Heisenberg's indeterminacy relations.

Contrary to Einstein's ideal of a complete (at least in principle) theoretical knowledge, a distinction is thus introduced between pure theory, expressed in a logically consistent (complete) language and the description of experience, an observational language which is, at times, "repugnant" to pure theory, i.e., inconsistent with it.

While this premise brings Schrödinger close to Bohr, a marked difference nonetheless remains between their respective positions, as seen in their divergent conclusions. Instead of insisting on the construction of a unitary theoretical language, Bohr prefers to adapt language to experience, at the cost of using two split classical languages [6] (complementarity). Schrödinger, instead, maintains that a consistent (complete) classical language is an indispensable guide, even if it speaks about itself and not the real world. In fact, even so, it contributes to knowledge of the world. He was convinced, in fact, that a true-knowledge should be comprehensive of all the contributions of culture, humanities included.[7]

While Einstein would have considered that, after Quantum Mechanics, science was confronted with a dramatic choice between total salvation (i.e. the rejection of the Copenhagen interpretation) or total destruction, Schrödinger proposed what he wittily called "the emergency exit," i.e., his conception of a two-level theory.

Against the Cophenhagenists, Schrödinger's position implies that renunciation of both COP and CAC at the purely theoretical level would unnecessarily destroy the very conditions for a good theorisation. If COP were rejected, something would be lost, not in knowledge of objective reality, as Einstein would have maintained, but in its adequacy, the grasp of theory on phenomena, its capability of disclosing wider areas of experience. In Schrödinger's continuous theory of micro-physics there are no gaps, on the purely theoretical level, both as regards picture (i.e., ST description) and causation; gaps in causation and description are however possible on the observational level.

However, when he added the further qualification that a contradiction might exist between the two levels, he stretched their independence to its extreme, introducing a quasi-dichotomy between pure theory and observational language.

This extreme position was not acceptable to the majority of Schrödinger's contemporaries and to Einstein in particular. Causal gaps (the breaking of CAP), even if limited to the observational level, could not be accepted by Einstein and other scientists who placed COP and CAC on the same conceptual level.[8] In fact, Einstein's completeness implied a bivocal correspondence between concepts and observables.

It followed from Einstein's premises that, if Schrödinger's wave function did not correspond to a complete description of the system, the reason was to be sought in its statistical (in Einstein's sense!) features: i.e. wave function refers to an ensemble of systems not to an individual system[9]. By contrast, Schrödinger thought that incompleteness in description was generated by an illegitimate (due to indistinguishability) individualisation of classical or quasi-classical particles in micro-physics.

On the other hand, Schrödinger could not accept Heisenberg's and Bohr's Copenhagenism because, for him, their position represented a concession to the old conception of the theory-observation relation, implying that causality-gaps and discontinuities on the observation-level would forbid the construction of a complete theoretical language. In fact, Bohr postulated the necessity of using two complementary languages[10] (his complementarity principle), when he found that one descriptive language alone was inadequate.

One can thus argue that Schrödinger considered the fundamental defect of the Copenhagen view to be that it missed the distinction between the two levels of language, the descriptive and the purely theoretical level. From the QM impossibility of a continuous descriptive language on the observational level, the Copenhagenists would have rushed to conclude that a continuous purely theoretical language was useless.

In the foregoing we have witnessed physicists' remarkable range of ideas in their effort to rationalise the world of perceptions. In this process, Einstein's argument about the lack of completeness of his GR represented a pivotal point. As seen above, he considered incomplete a theory in which the empirical content had to be introduced from without through a CCr, rather than being defined though the conceptual framework of the same theoretical system. This incompleteness was at bottom the manifestation of a failure in the synthesis between the formal a-priori and the intuitive content of experience, in which, according to Kant, consisted the process of an objective knowledge in CP.

Bohr completely changed the classical approach, thus bypassing rather than solving Einsten's problem above. Symbolisation was the intellectual activity at the root of his new conception.

In his proposal of a distinction/opposition between a purely theoretical and an observational language, Schrödinger was aware of the above problem. For him, scientists should renounce the a-priori synthesis, i.e., the Kantian presupposition of the objectivity of the physical world. In the lateSchrödinger this objectivity is rejected in favour of a science of processes, e.g., a transformation of energy-states, which is mathematically framed through the second-quantisation procedures.

I want to argue that these physicists' philosophical speculations were not brought into science from without as a piece of cultural entertainment for physicists who had lost touch with actual research. Quite on the contrary, philosophical problems forced their way into physics as an inner necessity, at times even against scientists' wish to keep their inquiry within the bounds of an empirical approach. In my foregoing essays, I have given evidence for the view that physicists must maintain a critical attitude towards the generalisation of theories and the theory-experiment relation.

The historical cases I have examined point to the existence of a connection between the specific development of TP and physicists' reflective "criticism". Evidence for this view is also given by the mere observation that, at the turn of the nineteenth century, philosophical speculation grew to an unprecedented degree, keeping pace with the exponential development of TP in the same period. As my reconstruction has indicated, TP emerged from a complex mixture of empiricial work and reflective criticism, a complexity which, in my opinion, is not adequately dealt with in most of the traditional histories of physics.

Acknowledgements

I gratefully acknowledge the inspiration and help I have received from the work of historians and philosophers of physics as quoted in my notes and bibliography. I am especially grateful to my friend and colleague Arcangelo Rossi for his perceptive comments on various topics closely related to epistemology and history of physics.

Roma, July 3, 1998

NOTES

Introduction

[1] Boltzmann [1905] 94; quoted in Jungnickel & McCormmach [1986] Vol. 1, Preface, xix.

[2] Wilhelm Wien, "Ziele und methoden der theoretischen Physik", *Jahrbook der Relativität und Elektronik*, 12 (1915) 241-59, 241; reported from: Jungnickel & McCormmach [1986] Vol.1, Preface, xviii.

[3] In France, Fourier and Ampère inspired a trend in mathematical physics, which had as its adepts Poisson and Chauchy, among others. Comte's philosophy somehow expressed the foundational conceptions of this trend, which also had as its roots the teaching and research of the Ecole Polytechnique .

[4] The institutional and conceptual aspects of German theoretical physics were masterfully investigated in Jungnickel & McCormmach [1986].

[5] Holton [1988] 31-147.

[6] This is the subtitle of Vol.1 of Jungnickel & McCormmach [1986].

[7] In Holton's historiography, mathematisation represents one of the dimensions of what Holton labelled the x-y contingent plane of physical research.

[8] Max Jammer, "A Consideration on the Philosophical Implications of the New Physics", in: Radnitsky & Anderson [1979] 41-62, 41.

[9] Cassirer [1950] 82.

[10] In this work, I use philosophy in a large sense of the word, without implying that physicists necessarily adopted a specific philosophical system. In pages dedicated to the thematic analysis of Newton's *Principia* and Einstein's special relativity, Holton treated this topic under the heading of the thematic dimension of scientific thought (Holton [1988] 31-147).

[11] Cassirer [1950] 82.

[12] H.Helmholtz [1878]; Helmoltz , *Vorträge und Reden*, vol II, pp. 215-247, 387-406; Cohen & Helkana [1977] 117.

[13] B. Bertotti [1987] *Presentazione*, 8. Also, Bertotti [1985]. A similar view is presented by Wessels [1983].

Chapter 1

[1] Ampère [1826] 10.
Ampère [1921] 79.

[2] This short rendering of Ampère's arguments is in Whittaker [1951] vol.I, 85, 86.

[3] When applied to natural magnets or lodestones, this challenging idea clashed with the difficulty of accepting that currents in the microscopic world could flow continuously without dissipation of heat.

[4] Ampère [1921] 79.

[5] Ampère [1921] 79, note.

[6] Ampère [1926] 161-194.

[7] Ampère [1926] 8.

[8] O'Rahilly [1965] Vol. 2, 523.

[9] This is a view held by both Ampère and von Humbol. See Jungnickel & Mccormmach [1986] Vol. 1, 64.

[10] A letter from Gauss to the Göttingen University Curator, 29 January 1833, a quotation in Jungnickel & McCormmach [1986] Vol. 1, 64.

[11] Gauss [1867] Vol. V, pp. 293-304.

[12] For a very incisive overview of Gauss's theoretical and experimental contribution to magnetism: Jungnickel & McCormmach [1986] Vol. 1, 63-77.

[13] Weber [1893] b) 6-18.

[14] A rather extensive discussion of Weber's life and work is in: Jungnickel & McCormmach [1986] a) 130-148; b) 72-79.

[15] Obviously, the inclusion in a wire of particles of both signs was to account for the well known fact that conducting wires are statically neutral.

[16] Weber [1846].

[17] Weber [1846] Introduction, 34.

[18] Weber [1846] 35-40. Therein Weber described this instrument in great detail.

[19] Weber [1846] 79.

[20] Weber [1846] 115 ff.

[21] Weber [1846] 112 ff.

[22] Weber [1846] 115 ff.

[23] Weber [1846] 103 ff.

[24] Weber [1846] 27-30, 133.

[25] Weber [1846] 132-134.

[26] Weber [1846] 135: "Diese drei Tatsachen sind zwar nicht unmittelbar durch die Erfahrung gegeben...sie hängen aber nit unmittelbar beobachteten Tatsachen so

genau zusammen, daß sie fast gleiche Geltung als sie haben".

[27] Weber [1846] 134-135.

[28] Weber [1846] 142.

[29] Weber [1846] 54 ff. A modern rendering of Weber's deduction is in Whittaker [1951] Vol. 1, pp. 83-88.

[30] Weber [1846] 140-44.

[31] Weber [1846] 44.

[32] Weber [1846]152; Weber [1851]; Weber [1848]; Weber [1893] b); Weber [1852].

[33] Weber [1846] ; Weber [1893] b) 157.

[34] Weber [1848]; Weber [1893] b) 245-246.

[35] Weber [1893] b) 157- 207.

[36] Weber [1851]; Weber [1893] b).

[37] Weber [1852] ; Weber [1893] b) 301-465.

[38] Weber [1893] b) 321.

[39] Weber [1846] 211.

[40] Weber [1848] 245.

[41] For clarity's sake, in order to account for the new unit of time, I use t^* in place of Weber's original notation t.

[42] Weber [1852] 366.

[43] Weber [1852] 366-67.

[44] Weber [1852] 358.

[45] Weber [1852] 368: "...ohne Kentnis der Geschwindigkeit die Reduktion der Gemessen Stromintensitäten, elektromotorischen Kräfte und Widerstände auf the bekannten Maasse der Mechanik nicht ausgeführt werden kann".

[46] Weber [1852] 368.

[47] Weber & Kohlrausch [1856] 604-05.

[48] Kohlrausch spoke of this experiment in the manuscript of a report to the Natural Science Society of Marburg dated 16 June 1852 [N.S.F. source] 74). This method was the one tried by Rowland in his well-known experiment. See also Jungnickel & Mc Cormmack [1986] a) 145.

[49] In 1873 Maxwell thought that the necessary velocity to obtain measurable effects was practically achievable (D'Agostino [1996] 41-42).

[50] Weber, "*Vorwort bei der öbergabe der Abhandlung: Elektrod. Maasbest. insbesondere Zurückfürung der Strömintensituats-Messungen auf Mechanische Maass*" in: Weber [1856] 592, ff .
Weber & Koholrausch, "*Elektrodynamische Maasbestimmungen insbesondere

Zurückfürung der Strömintensitäts-Messungen auf Mechanische Maass." Submitted by Weber to the Saxon Soc. in 1855, in Weber [1893] b) 609-76.

[51] Short versions of Weber's argument are often reported in the secondary sources; for instance: Whittaker [1951] Vol 1, 201-208. Also in : Wiederkehr [1967] 140-141. Therefore I can be excused from reporting the full procedure.

[52] Actually, the measurement of Q with the requested precision, implied the measurement of a known fraction of it, transfered to a large sphere of diameter 16 cm, and the usage of a *sine-electrometer,* an instrument invented by Koholrausch, who was especially expert in static measurements. Its usage required skilled craftsmanship.

[53] Weber and Kohlrausch could have used the simpler operations with the electrodynamometer for this purpose. However, due to the higher sensitivity requested by their procedure, they preferred the more complicated and less indirect operations of the *ballistic galvanometer,* also relying on Weber's long prior experience with Gauss in makingabsolute magnetic and electromagnetic measurements. The coil of the instrument used by Weber and Kohlrausch in the main version of their 1857 experiment contained 5635 turns of mean diameter 26.6 cm, i.e., some 4.7 km of wire; probably a record for its day (NSF source).

[54] Weber & Koholrausch [1856] 605.

[55] Weber [1856] 595.

[56] Kirchhoff, "*Über die Bewegung der Elektrizität in Drähten*", [1857], in: Kirchhoff [1882] 131.

[57] Weber, "*Elektrtrodynamischen Massbestimmungen insbesondere elektrische Schwingungen*", in: Weber [1864].

[58] Weber [1864] 157.

[59] Weber & Koholrausch [1856] 607.

[60] Weber & Koholrausch [1856] 607.

[61] The same consideration is true in general concerning Weber's measurement .

[62] Maxwell, highly praises Ampère's theory in is *Treatise*. See: Maxwell [1954] Part IV, Chap. III.

[63] Weber's remarks on Ampère are extensively quoted by Duhem [1921] 324.

[64] Duhem [1921] 324.

[65] Duhem [1921] 322.

[66] Ampère [1826] 5.

[67] Notice that this mode of relating symbols and experimental data is subjected to the constraint that quantities be represented by real numbers times a unit determined by instrumental operations. This form of contraint is overcome in quantum mechanics where physics laws are represented via operators including immaginary numbers.

Chapter 2

1 Fourier [1822].

2 Fourier [1822] Art. 161.

3 Maxwell [1954] Vol. I, Art.42.

4 Kelvin [1981] 104, quoted by O'Rahilly [1965] Vol. 2, 697.

5 Clifford [1878] 49, quoted by O'Rahilly [1965] Vol. 2, 685. O' Rahilly's opinion on dimensional quantities in physics is unambiguously expressed : "The mystification sets in when we begin to misinterpret these numbers as complex qualitative happenings miraculously susceptible to arithmetical operations such as raising to the forth power. It is precisely the failure to recognise the symbols of physics as ordinary numbers, which has led electricians into such a quagmire of futile and meaningless methaphysics" (O'Rahilly [1965] vol. 2). See also:Crowe [1967].

6 Clausius [1982].

7 Clausius [1982] 536.

8 Clausius [1982] 543,544.

9 Planck [1932], *Theory of Electricity and Magnetism*, vol. III.

10 M. Loria, "Giovanni Giorgi", in: Gillispie [1972] vol. 5, 407.

11 G. Giorgi, "Unità (sistemi di)" in: *Enciclopedia Italiana* [1937] vol. 34, 714-718, 716.

12 Sommerfeld [1935] (the passage was translated into English by the present author).

13 Sommerfeld [1964].

Chapter 3

1 Weber, "Elektrodynamische Maasbestimmungen. Über ein allgemeines Grundgesetz der Elektrischen Wirkung" [1846] in: Weber [1893] 25-211.

2 Weber & Kohlrausch [1856] 597-608.

3 Among them Bromberg [1957]; Everitt [1975].

4 D'Agostino [1980].

5 Larmor [1973] 705.

6 Larmor [1973] 729.

7 Because of Weber's choice of electrodynamic units the factor 1/2 appeare before the velocity c. However Weber adopted also electromagnetic units (D'Agostino [1980] 285).

⁸ Maxwell, "On Physical Lines of Force", in: Maxwell [1861]; Maxwell [1854], Vol. I, 451-513.

⁹ Maxwell [1854] 500.

¹⁰ On this point: Bromberg [1957] 227; Heimann [1870] 193. Both authors agree on the fact that Maxwell did not know Weber's numerical value when he discovered that the velocity of his magnetic waves was equal to Weber's factor. In the following, he would have discovered that this factor was equal to the velocity of light. For the argument of my paper is irrelevant whether Maxwell's identification of the velocity of electromagnetic waves with that of light was or was not a discovery.

¹¹ Maxwell [1850].

¹² Maxwell [1861] 495.

¹³ Maxwell [1861] 498.

¹⁴ Maxwell [1861] 500.

¹⁵ Niven [1980] xvi.

¹⁶ Niven [1980] xvi.

¹⁷ "Report of the Committee appointed by the British Association on Standards of Electrical Resistance", in: Maxwell & Jenkin [1863] 111-176.

¹⁸ Although I do not deal in this work with Thomson's program of precise measurements, I wish to call attention to this important section of Thomson's research, one that was recently studied in great detail by Crosbie Smith and Northon Wise (Smith & Wise [1989]).

¹⁹ Maxwell & Fleming [1863] 130-163, 131.

²⁰ Maxwell & Fleming [1863] 130-163, 132.

²¹ Maxwell & Fleming [1863] 130-163, 132-135.

²² The relevance of Fourier's analysis for Maxwell and Thomson's measurement program is duly underlined in Smith & Wise [1989] 125,150, 161, passim...

²³ "The Report of the Committee"[1893], 118.

²⁴ Equation (8) is the above : $R=\sqrt{\dfrac{VSL}{C}}$.

²⁵ Larmor [1907] 26.

²⁶ C.W.F. Everitt valuates highly this unduly neglected paper by Maxwell because it "supplied a vital step" in the definition of a "dual system of electrical units " (Everitt [1975] 100). I agree with Everitt on this point. He continues by stating that "by 1863, then, Maxwell had found a new link of a purely phenomenological kind between electromagnetic quantities and the velocity of light" (ibid. 101). I think that the construction of the dual system, from which Maxwell's new link was derived, is far from being phenomenological, although I admit that the new link requires a minor number of ad hoc hypotheses in comparison with physical lines. For

other aspects of the 1863 "Report": D'Agostino (1978).

[27] Maxwell [1864].

[28] Maxwell [1864] 536,568.

[29] Maxwell [1864] 569.

[30] Maxwell [1864] 579.

[31] Maxwell [1864] 563-564.

[32] Maxwell [1864] 589-597.

[33] Maxwell [1868].

[34] Maxwell [1868] 128.

[35] Note that two systems of units were used consistently in the same equations (the conversion factor was introduced to satisfy an homogeneity principle)

[36] Maxwell [1868] 134-135.

[37] Maxwell [1868] 143.

[38] Maxwell [186]) 134-135.

[39] However, Maxwell's above deduction lacks the generality of Weber's deduction of his factor as presented by Weber in his law of force. From this deduction the factor derives its full theoretical justification.

[40] Maxwell [1891]; Maxwell[(1954].

[41] Entitled "Preliminary".

[42] Maxwell [1954] §.2.

[43] Maxwell [1954] § 625.

[44] Maxwell [1954] § 624.

[45] Maxwell [1954] §§ 625, 686.

[46] Maxwell [1954] § 526.

[47] Maxwell [1954] §§ 620-629

[48] Maxwell [1954] § 622.

[49] Maxwell [1954] § 624

[50] Maxwell [1954] § 626.

[51] Maxwell [1954] § 627.

[52] Maxwell [1954] § 627.

[53] Maxwell [1954] § 628.

[54] Maxwell [1954] § 768.

[55] Maxwell [1954] §§ 768-770.

[56] This system Maxwell had analysed in detail in § 653 ff.

[57] Maxwell [1954] § 769.

[58] A note added by J.J. Thomson informs us that the effect was discovered by Rowland in 1876.

[59] I think that Maxwell's above distinction between a classification of v as a real quantity in the first conceptual experiment and as a physical quantity in the second may be deepened by analysing Maxwell's ideas on a physical classification of quantities as distinct from their mathematical classification, a point that he makes in his Lecture "On the Mathematical Classification of Physical Quantities" (Maxwell [1954] a) Vol. 2, 257-266.

[60] See Smith & Wise [1989].

[61] Maxwell [1954] §§ 781-805.

[62] Maxwell [1954] § 786.

[63] Eight experiments of various types to measure the ratio of units, some of them performed by Kelvin, others proposed or performed by Maxwell himself(Maxwell [1954] §§ 772-80)

[64] Maxwell [1954] § 787.

[65] Maxwell [1954] § 786.

[66] Smith & Wise [1989].

[67] Schaffer [1994].

[68] I comment on this matter in D'Agostino [1986] 194-198.

[69] Schaffer [1994] 139.

[70] Schaffer [1994] 146 ff. Also in: Smith & Wise [1989] 455.

[71] Raleigh [1915]. Other remarks in :Carneiro [1993].

[72] [Buckingham] 19XX. Carneiro [1993].

[73] These was the form in which physical laws were expressed in the writings of Coulomb, Fresnel and others physicists of the beginning of the nineteenth century.

[74] D'Agostino [1996]46-49.

[75] Notice that this mode of relating symbols and experimental data is subjected to the constraint that quantities be represented by real numbers times a unit determined by instrumental operations. This form of constraint is overcome in quantum mechanics if physical laws are represented by operators including imaginary numbers.

[76] Let me express my agreement on this point with Smith and Wise. In commenting on Kelvin's statement that, in order to meet a scientific test, any quantity was to be measurable, they keenly remark that "Maxwell's displacement current ...had never been observed, let alone measured in the sense direct sense Thomson intended" (Smith & Wise [1989] 455).

[77] Maxwell [1954] § 526.

[78] D'Agostino [1996] 47. In my paper I gave evidence of the fact that Maxwell

never quoted or utilized Weber and Koholraush's velocity c, but in his research he limited his approach to Weber's ratio of units (see above Gauss's and Weber's metrology). On this point, see also Siegel [1991] 130.

[79] Sommerfeld 1935], [1964] 53-54.

[80] Panofsky & Phillips [1955] 375-378.

[81] However, the authors add that Maxwell's choice of only three fundamental mechanical units does not allow a numerical determination of the absolute values for the ethereal constants. This determination is possible if a fourth non mechanical unit is arbitrarily determined Panofsky & Phillips [1955] 376.

[82] Panofsky & Phillips [1955] 375.

[83] Panofsky and Phillips mantain (Panofsky & Phillips [1955] 376) that c in the expression above was first determined by Weber and Koholrausch by measuring the discharge of a condenser whose electrostatic capacity was known. In my study I proved that this velocity was not Maxwell's velocity c (v in Maxwell's and my notation).See D'Agostino [1996] 47.

[84] As it can be also argued by referring to Buchwald's very detailed analysis in: Buchwald [1985] 27-33, 37-40, 47.

[85] Buchwald [1985] 23 ff.

[86] Notice that had Maxwell decided to write (as Weber did) $ids = ev$, he could have easily proved that the ratio of units = velocity C.

[87] Weber's expression is evidently equivalent to the modern local expression :
$\rho_{esu} u = J_{esu}$.

[88] In Einstein's relativity, when F_d and F_s stand for the Coulomb and Ampère forces, one has: $F_d/F_s = (v/C)^2$, v convection velocity of charge, C velocity of light. See for instance: Feynmann [1965]. If C is identified with Weber's c, the same relationship holds in Weber's theory.

Chapter 4

[1] Albert Einstein clearly recognised in various passages the revolutionary import of Maxwell's theory (Einstein [1949]c) 33).

[2] A clear understanding of Maxwell's revolutionary view of the relationship between mathematics and physics can be grasped by a comparison between his and Thomson's views as underlined by Smith and Wise. Thomson went out to slay the nihilist infidels of mathematical as opposed to physical ideas (Smith & Wise [1989] 445).

[3] Maxwell [1855] 187.

[4] Maxwell [1855] 208

5 Maxwell [1855] 208

6 D'Agostino [1968].

7 Boltzmann,"On the development of methods of theoretical physics" [1892], in Boltzmann [1974] 77-100, 89.

8 Poincaré [1890], Introduction, p. IV.

9 Maxwell [1891] [1954].

10 Maxwell [1891] [1954] Part IV, Chap.V, § 553, 199.

11 Maxwell [1891] [1954] Part IV § 554, 199.

12 Maxwell [1891] [1954] Part IV, Chap. XX1, § 818.

13 D'Agostino [1968] b.

14 Maxwell [1891] [1954] Part IV, , Chap. V, § 554, 209. In his review paper, dealing extensively with Maxwell's dynamical ideas, Norton Wise refers to G.G. Stokes' distinction between "mechanical" and "dynamical" theories, the latter consisting in a methods that do not require the physicist "to assume either that the ether does or that it does not consist of distinct particles" (Wise [1992] 186).

15 Buchwald [1985], XII. As remarked by Buchwald, historians, physicists and philosophers used to interpret Maxwell's method as a method mainly based on analogies, thus missing what was typical in Maxwell's DA.

16 Maxwell [1891] [1954] Part IV, Chap. IV, § 552, 198.

17 Maxwell [189]) [1954] Part IV, Chap. V, § 567, 200.

18 Maxwell [1891] [1954] Part IV, Chap. IX, § 604, 247. According to Buchwald, the same ideas were shared by the Maxwellian, the physicists who pursued Maxwell's approach (Buchwald [1985] 96 ff.).

19 Buchwald [198]) 27-33, 37-40, 47.

20 In symbols, this would be represented in the well known equation:

$\rho_{esu} u = J_{esu}$ (or in the similar Weber's convective conception: $i = m e^* u$). The apparently similar expression adopted by Maxwell, $u = sv$, (in es units) concerned the motion of macroscopic charged bodies and represented a conceptual experiment of the Rowland-type (On this point: D'Agostino [1996] 5-51.

21 Maxwell [1954] a), Vol .2, 362.

22 I agree with Wise when he complains that Maxwell's dynamical ideas have not been taken into account by his commentators in discussing his views on gas theory. In this chapter composed before reading Wise's work, I have complied with Wise's requirements. In particular, my thesis agrees with his statement that Maxwell's most sophisticated treatment of kinetic theory bears close comparison with his Lagrangian treatment of electrodynamics. I also agree with Wise's remark on Maxwell's and Thomson's special view of molecular physics. Although both recognised the necessity of a molecular physics "they also treated molecules macroscopically, as stable mechanical systems whose structure they did not pretend to

specify. This does not imply that Maxwell completely adhered to Stokes's and Thomson's positions (Wise [1992] part 1, 191, 185, 196-198).

[23] Maxwell, "On the Dynamical Evidence of the Molecular Constitution of Bodies", in: Maxwell [1954] a), Vol. 2, 418.

[24] Maxwell, "Atom", in *The Encyclopaedia Britannica*, and in Maxwell [1954] a), 445-84.

[25] Maxwell [1954] a) 470-471.

[26] Maxwell [1954] a) 471.

[27] Maxwell [1954] a) Vol. 2, 215-229.

[28] Maxwell [1954] a) Vol. 2, 223-224.

[29] Maxwell [1954] a) Vol. 2, 471.

[30] Maxwell, " On Stresses in Rarefied Gases arising from Inequalities in Temperature", Maxwell [1954] a) Vol. 2, 681-712. Maxwell died in November of the same year.

[31] Maxwell, Appendix to "On stresses in Rarefied Gases", added in May 1879, cited in: Maxwell [1954] a) Vol. 2, 703-712, 704.

[32] Maxwell, " On Stresses in Rarefied Gases Arising from Inequalities in Temperature", Maxwell [1954] a) Vol. 2, 681-712, 681.

[33] Maxwell [1877], Maxwell [undated] Dover Publ.

[34] Maxwell [undated] 122.

[35] Clearly, Maxwell's position here prefigures an ante-literam Popperian falsificationism .

[36] Maxwell [undated] Dover Publ., 122.

[37] The above conclusion may explain why, in Maxwell's theory, one can clearly distinguish three different approaches to thermodynamics: one is described as the purer thermodynamics, epitomized by the phenomenological Clausius's theory, to which Maxwell himself gave important contributions, such as the distribution law of velocities and the equipartition of energy. Another approach was described as purely dynamical, and he referred to some attempts of Clausius at a pure dynamical theory of thermodynamics, the object of his criticism above. He quoted a few of Clausius' papers as examples thereof. In his paper "Tait's Thermodynamics" he affirmed that the attempts to deduce the second law from purely dynamical principles were unsuccessful because they led to unsound results (he quotes as an example Clausius [1872]). One may infer that his criticism was also referred to Boltzmann's attempts at a mechanical explanation of the second law. The third approach was inspired by his conception of DA, and he labelled it a statistical approach.

[38] As Harmann very appropriately remarks in his Introduction to Maxwell's Scientific Papers and Letters, "Maxwell's disjunction between the laws of thermodynamics and the laws of mechanics led him to be severely critical of the attempts,

notably by Clausius and Boltzmann, to reduce the second law of thermodynamics to a theorem in dynamics" (Harmann [1995] 18).

[39] Maxwell [undated] Dover Publ.,122.

[40] Helmholtz [1878]; Helmholtz [1977].

[41] Cappelletti [1976].

[42] Helmholtz [1977] 118.

[43] Helmholtz [1977] 119.

[44] Helmholtz [1977] 122.

[45] Helmholtz [1977] 122.

[46] Helmholtz [1977] 118, 119.

[47] Helmholtz [1977] 162.

[48] For Helmholtz the general form is that schema devoid of any content which he declared "to be the true form of intuition, in respect of which Kant's doctrine of the a-priori is to be upheld"; (M. Schlick, notes to *Hermann von Helmholtz Epistemological Writings*, in : Helmholtz [1977] 172, note 33).

[49] Helmholtz [1977] 162.

[50] Helmholtz [1977] 121. Helmholtz compared the different perceptive characteristics of eye and ear and concluded that the former's incapability to detect beat frequencies represented its modality: "could the optical nerve at all follow in sensations the enormously rapid beats of light oscillations, then every mixed colour would act as a dissonance".

[51] Helmholtz [1977], appendix n 3, 163.

[52] Helmholtz [1977] 117.

[53] Helmholtz [1977] 141.

[54] Helmholtz [1977] 141.

[55] M. Schlick, notes to *Hermann von Helmholtz Epistemological Writings*, in : Helmholtz [1977] 172.

[56] D'Agostino [1990] 391.

[57] Hertz [1894], Hertz [1956] Author's Preface (pages not numbered).

[58] Hertz [1956] 1.

[59] Hertz [1956] 2.

[60] Hertz [1956] 25.

[61] Hertz [1956] 2.

[62] Hertz [1956] 23.

[63] Hertz [1956] 45.

[64] Hertz [1956] 45.

⁶⁵ I argue that Hertz's conception of an independent or, in modern terms, self-sustaining force in ether, in the form of ether polarization, and its propagation, expressed at its best the truly innovative idea of a field theory and contiguous propagation at the end of the nineteenth century. The next step, which Einstein achieved in his general relativity theory, was the attribution of a structure, not to a matter-type substance diffused in space, but to space-time itself. The theory confirmed, among other things, the fecundity of the abolition of the general concept of a source of forces, the abolition Hertz had so early adopted in his electrodynamical researches (See: D'Agostino [1993] a)).

⁶⁶ Hertz [1891], Hertz [196]).

⁶⁷ D'Agostino [1993] a).

⁶⁸ Boltzmann, "On the Fundamental Principles and Equations of Mechanics" (1899), in: Boltzmann [1974] 119.

⁶⁹ D'Agostino [1989].

⁷⁰ Hertz, "The Forces of electric Oscillations", in: Hertz [1962] 138 .

⁷¹ The abolition of sources and of the classical concept of the force-source connection was certainly a paradigm shift of the utmost significance, as can be argued by the resistance it met among traditional physicists. As an example of this resistance, even at a later date, I like to quote Max Laue's objection to Einstein's equivalence principle in 1911, when have argues that a gravitational field produced by accelerating frames cannot be real since it has no masses as its sources; in: Einstein,"Dialog über Einvände gegen die Relativität Theorie", *Die Natürwis 6* [1918] 700. Quoted in: Norton [1985]. See also: D'Agostino [1993] a).

⁷² D'Agostino [1990].

⁷³ Hertz [1962] 159.

⁷⁴ Hertz [1962] 3. On this point, see also: Helmholtz's Preface to Hertz [1956] xi.

⁷⁵ Hertz [1962] 197.

⁷⁶ Hertz [1962] 197. Notice that Hertz uses correctness, not truth, concerning the probation value of an experimental confirmation of a theory.

⁷⁷ Hertz [1962] 197. This conception of the theory-experiment relationship was to be named later as the Duhem-Quine hypothesis.

⁷⁸ Hertz "On Electromagnetic Waves in Air and their Reflections", in: Hertz [1962] 136

⁷⁹ Hertz, "The forces of Electric Oscillations Treated According to Maxwell's Theory", in : Hertz [1962] 159. Poincaré argued that Hertz's experiments were not crucial for Maxwell's theory. The French scientist believed that a true theory should be crucially tested by an experiment. On this point: D'Agostino [1986].

⁸⁰ This partial autonomy between experiment and theory somehow justifies Buchvald's conclusion that "Hertz's experiments present a peculiar character

which differentiates them from others such as Fresnel's or Hall's equally important experiments...Hertz's experimenting...was designed to show that something does not occur or else to find something new that was not required by the kind of physical scheme that Hertz deployed. Hertz's experimental work, unlike theirs, had little to do with theory." (Buchwald [1994], Preface xiii).
In my view, Hertz's experimental work prefigured a new type of relation between theory and experiment, one which was chracterized by Holism and theoretical pluralism.

[81] The problem of a modern view of the theory-experiment relation that sheds light on Hertz's ideas is analysed by C. Chevalley in her recent:"Le conflict de 1926 entre Bohr et Schrödinger: un example de sous-determination des théories" (A lecture to the XIX International Congress of History & Philosophy of Science; Chevalley [1991]).
Cassirer interpreted Hertz's ideas in terms of his neo-Kantian philosophy: "Thus the individual concept can never be measured and confirmed by experience for itself alone, but it gains this confirmation always as a member of a theoretical complex...No element can be separated from the total organism and be represented and proved in this isolation. We do not have physical concepts and physical facts in pure separation, so that we could select a member of the first sphere and enquire whether it possessed a copy in the second; but we possess the "facts" only by virtue of the totality of concepts, just as, on the other hand, we conceive the concepts only with reference to the totality of possible experience....Here the function of the concept only extends to the subsequent inclusion and representation of the empirical material; but not to the testing and proving of this material....Those thinkers, also, who strongly urge that experience in its totality forms the highest and ultimate authority for all physical theory, repudiate the naive Baconian thought of the *experimentum crucis*". (Cassirer [1923] 147).
See also: Cassirer [1950] 106 ff.

[82] Mach [1926] 2.

[83] Helmholtz [1878] 170.

[84] Mach, *Analyse der Empfindungen*, in : Schlick's notes to Helmholtz (1878) 177.

[85] Helmholtz [1878] 172.

[86] Mach [1894]; Mach [1943] 241.

[87] Mach [1943] 253.

[88] Mach [1943] 253.

[89] Mach [1943] 248.

[90] Mach [1943] 248.

[91] Mach [1943] 177.

[92] Mach [1943] 253.

[93] Using a well-known piece of Kantian terminology, it can be said that theory has

for Mach just a *regulative* import as regards occurences and that it thus serves as a guideline for inquiry, but, concerning its validity, it cannot claim any further guarantee than its same success.

94 Cassirer [1950] 84.

Foreword to Part Two

1 I mean the theory Hertz fully developed in his theoretical paper: "On the Fundamental Equations of Electromagnetics for Bodies at Rest", Hertz [1962] 195-240.

2 Planck [1894] 278.

3 Planck [1894] 278; Planck [1931] 356; a quotation in Jungnickel & McCormmach [1986] Vol .2, 47. Planck's remarks were unknown to me at the time of my 1975 work.

Chapter 5

1 Hoppe [1928], 553.

2 Smith & Wise [1989] 446, 661.

3 Whittaker [1951], 206.

4 Helmholtz [1882] *Über die Erhaltung der Kraft. Eine physikalische Abhandlung*, 12-75, 62.

5 Whittaker [1951], 218.

6 Hoppe [192]), 590-592; Whittaker [1951], 203.

7 Helmholtz [1882] *Über die Theorie der Elektrodynamik*, 634-645, 639.

8 «Die entstehenden und vergehenden Componenten dieser Polarisation wurden den Strom constituiren der durch das astatische Nadelpaar angezeigt wird» (Helmholtz [1882], 797). Thomson J.J. [1881] attributed the magnetic effect of moving electrostatic charges to the continuous alteration of the electric field in the surrounding medium, or, in the language of Maxwell, to the displacement current. Whittaker [1951], 306-307.

9 Helmholtz [1869] 429, 531.

10 Rosenfeld [1956] 1636-1637. D'Agostino [1996] 1-51.

11 Simpson [1966] 411-413, 416.

12 Maxwell [1954] Vol. 1, 451-452 (§ 326), 487-488 (§ 355).

13 Maxwell [1954] Vol.1, 438-440 (§ 790).

14 Maxwell [1954] Vol. 2, 139.

15 Susskind [1964] 32-42.

[16] Whittaker [1951] 323.

[17] Together with its capacitance, the inductance of a wire was considered the cause of propagation in the wire itself and the process of induction in space was fundamental to the free propagation of electric disturbances in Maxwell's theory.

[18] Smith & Wise[1989] 455-457. As an evidence of the remarkable contrast between Thomson's and Maxwell's views on the propagation in wires, it is worthy to notice that Thomson opposed his telegraph theory to Maxwell's equations as "measurable" is opposed to" metaphysical"(Smith & Wise [1989] 454).

[19] D'Agostino (1975) 273-276: Buchwald (1994) Appendix 13,"Propagation in Helmholtz's Electrodynamics", 385-388. See also: letter from G.F. Fitzgerald to W. Thomson, April 25, 1885 (Smith & Wise [1989] 459).

[20] Hertz, "From Herr Wilhelm von Bezold's Paper: Researches on the Electric Discharge-Preliminary Communication", in: Hertz [1962] 54-62, 55.

[21] Maxwell [1954] Vol.2, 448-449.

[22] Rosenfeld [1956] 1635.

[23] Woodruff [1962] 439-459.

[24] Heimann [1970] 171-213.

[25] Bromberg [1967].

[26] Helmholtz [1870] b).

[27] Helmholtz [1870] a) 567. In this paper I use a modern vectorial notation throughout in place of the original coordinate notation. Here vectors are indicated by the x-component in the original text. The meaning of the symbols is indicated in eacch section separately.

[28] Helmholtz [1870] a), 568.

[29] Helmholtz [1870] a), 611-612.

[30] Helmholtz [1870] a), 625.

[31] Helmholtz [1870] a); Helmholtz [1881] 724-725.

[32] Helmholtz [1870] a) 556.

[33] Helmholtz [1870] a) 614.

[34] Helmholtz [187]) a) 627.

[35] Elkana [1970] 282. Among others, Elkana examines Helmholtz's conception of the relation between force and matter.

[36] Helmholtz [1870] a) 556-557.

[37] Helmholtz [1881] 819-820.

[38] Poincaré [1890] Chapter 5.

[39] Duhem [1902] 225.

[40] Rosenfeld [1956] 1665.

[41] Koenigsberger [1965] 293.

[42] Helmholtz [1982] 629-635, 629.

[43] Helmholtz [1982] 634.

[44] Helmholtz [1982] 774-790, 780.

[45] Helmholtz [1982] 791-797, 791.

Chapter 6

[1] For details of Hertz's life, see "Hertz, Heinrich Rudolf", in: McCormmach [1971] Vol. 6, 340-349. Hertz's scientific formation is studied by Buchwald [1992], "The Training of German Research Physicist Heinrich Hertz", in: Nye, Richard and Stuwer [1992].

[2] Heinrich Hertz, "Research on the Determination of an Upper Limit for the Kinetic Energy of the Electric Current", in: Hertz [1880] 414-448; Hertz[1896] 1.

[3] Hertz [1896] 3.

[4] Hertz [1896] 33.

[5] Hertz [1896] 34.

[6] Helmholtz's Preface to Hertz [1956] xxviii.

[7] Hertz [1962] Introduction, "Theoretical" 23-26.

[8] Hertz [1962] 1.

[9] Hertz [1962] 2.

[10] "Im Laboratorium die Arbeit begonnen über schnelle Schwingungen". Note for 7 September 1887, in: Hertz Johanna [1928] 175.

[11] Heinrich Hertz, "On the Relations between Maxwell's Fundamental Electromagnetic Equations and the Fundamental Equations of the Opposing Electromagnetics", in Hertz [1884]; Hertz [1896] 273-290.

[12] Hertz to Helmholtz, 21 January 1888: "Bisher habe ich noch nicht versucht, eine bestimmte Theorie auf die Erscheinungen anzuwenden und etwa die Constante K oder die Dielektricitatsconstante des Raumes zu bestimmen. Nur glaube ich mich überzeugt zu haben, dass das Maxwellsche Gleichungssystem nicht genügt. Wenigsten vermochte ich aus demselben überhaupt keine bestimmte Geschwindigkeit in Drahten abzuleiten" (*Hertz's Correspondence*, Deutsches Museum, MS. 3120).

[13] According to Doncel (Doncel [1991]), these conclusions were not included in the Academy report and were added by Hertz on 17 February 1888, to the paper published in *Wiedemann Annalen*.

[14] Hertz [1962] 123, Paper No. 7, Conclusions.

[15] Hertz [1962] 2. Concerning the further development of his experiments, Hertz

maintained that "in altering the conditions I came upon the phenomenon of side sparks [secondary sparks] which formed the starting point of the following research".

[16] Hertz, "On Very Rapid Electric Oscillations", in: Hertz (1962) 29-54, 29.

[17] Hertz (1962) 33.

[18] In a letter to Helmholtz, dated December 1886, Hertz writes: "I take this opportunity of communicating certain experiments in which I have been successful, because I was in hopes when I undertook them that they might interest you. I have succeeded, unmistakably, in showing the inductive action of one open rectilinear current, and I venture to hope that this method will eventually yield the solution of one or other of the questions associated with this phenomenon" cited in: Königsberger [1965] 368.

[19] Although all of these aspects of high frequency oscillations propagation in the short wavelength region and open resonators are of course connected in contemporary theories, the connection between them was not so clear in the theories of the time, especially since the concepts of distributed capacity and induction had not been clearly formulated.

[20] Hertz, "On Very Rapid Electric Oscllations", in: Hertz [1962] 29-53, 42-43.

[21] Hertz, "On an Effect of Ultraviolet Light Upon the Electric Discharge", Hertz [1962] 63-79.

[22] To obtain the correct value, the half period T that Hertz first obtained should be multiplied by 1/2, yielding: 1.26×10^{-8} sec. The consequences of this error affected all of Hertz's subsequent calculations of T in his experiments on propagation both in wires and in air. (In the free propagation experiments, however, the error in T had no practical consequences, since it was of the same order of magnitude as the imprecision in the measurement of λ in a system of stationary electromagnetic waves).

[23] Henri Poincarè, *Comptes Rendus*, 3 [1891], 322.

[24] Hertz, "Supplementary Notes, 1891", in: Hertz [1962] 270.

[25] Hertz, "Introduction, A) Experimental", in: Hertz [1962] 4-5.

[26] Hertz, "On the Action of a Rectilinear Electric Oscllation Upon a Neighbouring Circuit", *Wiedemann's Annalen*, 34 [1888], 155; reprinted in: Hertz [1962] 80-94.

[27] Hertz [1962] 82-83.

[28] Hertz [1962] 90-91.

[29] Hertz [1962] 86-91.

[30] Poincarè disagreed with Hertz's interpretation of nodes: letter from Poincarè to Hertz, 11 September 1890 (Hertz's Correspondence, Deutsches Museum, MS. 3000).; A. Righi maintained that the gap corresponded to an antinode, and modern theories would not accept either Hertz's or Righi's extreme positions.

[31] Lodge [1894] 8-9.

[32] Hertz, "On Electromagnetic Effects Produced by Electrical Disturbances in Insulators", *Sitzungsber. d. Berl. Akad. d. Wiss.* (10 Novembre 1887); in *Wiedemann's Annalen*, 34 [1888], 273; reprinted in: Hertz [1962] 95-106.

[33] Hertz [1962] 96. D'Agostino [1976] 305.

[34] Hertz [1962] 101.

[35] Hertz [1962] 101.

[36] A detailed interpretation of Hertz's Academy Experiment in: Buchwald[1994] 254-261.

[37] Hertz, "On the Finite Velocity of Propagation of Electromagnetic Actions", presented at the Berlin Academy on 2 February 1888;reprinted in: Hertz [1962] 107-123.

[38] Hertz [1962] 107.

[39] Hertz [1962] 108.

[40] In the summer of 1888 Hertz ascribed to Heaviside and Poynting the understanding that "the electric force which determines the current is not propagated in the wire itself, but under all circumstances penetrates from without into the wire". He considered it as "the correct interpretation of Maxwell's equations as applied to this case".In: Heinrich Hertz, "On the Propagation of Electric Waves by Means of Wires", *Wiedemann's Annalen*, 37 [1889], 395; reprinted in: Hertz [1962] 160-171, 160-161.

[41] Hertz [1962] 121. Hertz [1962] 151.

[42] Hertz [1962] 121.

[43] Hertz,"On Electromagnetic Waves in Air and Their Reflection." *Wiedemann's Annalen*, 34 [1888], 610; reprinted in: Hertz [1962] 124-136.

[44] He attributed the small discrepancy to the imprecision in the measurement of wavelength. I think that his error in the computation of the capacity affected the computed value for the period.

[45] Hertz [1962] 133.

[46] Hertz, "The Forces of Electric Oscillations, Treated According to Maxwell's Theory", *Wiedemann's Annalen*, 36 [1889], 1; reprinted in: Hertz [1962] 137-159.

[47] Hertz [1962] 140.

[48] Hertz's theory of a source-field relation had an antecedent in a similar theory that G. Francis Fitzgerald communicated to the Royal Dublin Society in 1883. Fitzgerald solved the d'Alembertian equation for the vector potential of a small circular loop of current, and he computed the emitted energy as a function of frequency. In a short sequel communication he remarked that by discharging a condenser through a small resistance, waves of ten meters or less could be produced. See Fitzgerald [1902] 93.

[49] The dependence on the inverse cube of the wavelength in Hertz's formula for the

emitted energy is the same as that given by Fitzgerald in 1883.

[50] Hertz, *Wiedemann's Annalen*, 36 [188]), 1; reprinted in: Hertz [1962 150.

[51] Hertz [1962] 151.

[52] Hertz [1962] 154.

[53] Hertz [1962] 156.

[54] Hertz [1962] 159.

[55] Hertz [1962] 146.

[56] Hertz [1962] 146; .D'Agostino [1976] 318.

[57] Hertz, "On the Propagation of Electric Waves by Means of Wires", in: Hertz [1962] 160-171.

[58] Heinrich Hertz, "On the Mechanical Action of Electric Waves in Wires", *Wiedemann's Annalen*, 42 (1891), 407; reprinted in: Hertz [1962] 186-194.

[59] Hertz [1962] 160-161.

[60] In supplementary notes to Electric Waves, written in 1891, Hertz elaborated on his new conception. There he maintained that he had experimentally proven the following effect: "...in the case of rapid variations of currents the change ["die Veranderung"] penetrates from without into the wire. It is thereby made probable that in the case of a steady current as well, the disturbance ["der Vorgang"] in the wire itself is not, as has hitherto been assumed, the cause ["die Ursache"] of the phenomena in its neighbourhood; but that, on the contrary, the disturbances in the neighbourhood of the wire are the cause of the phenomena inside it. That the disturbances in the wire are connected with a regular circulation of material particles, or a fluid assumed ad hoc, is a hypothesis which is neither proved nor disproved by our experiments; they simply have nothing to do with it ["eine Hypothese ... auf welche sich unsere Versuche gar nicht beziehen"]. We have neither any right to oppose this hypothesis, nor have we any intention of doing so ["noch ist es unsere Absicht"], on the ground of the experiments here described (Hertz "Supplementary Notes, 1891", in: Hertz [1962] 269-278, 175, note 24).

[61] Hertz, "On Electric Radiation", Sitzungsber. d. Berl. Akad. d. Wiss. (13 December 1888); in *Wiedemann's Annalen*, 36 [1889], 769; reprinted in: Hertz [1962] 172-185.

[62] Hertz [1962] 182-183.

[63] Hertz [1962]124-136, 124.

[64] Hertz [1962] 136.

[65] Hertz, "On the Fundamental Equations of Electromagnetism for Bodies at Rest", in: Hertz [1962] 195-241, 236.

[66] Hertz "Introduction" 1-28, B) Theoretical, in: Hertz [1962] 22.

[67] Hertz, "Introduction, A) Experimental", in: Hertz [1962]. 6.

[68] Hertz, "Introduction, A) Experimental", in: Hertz [1962] 7.

[69] Hertz, "Introduction, A) Experimental", in: Hertz [1962] 7.

[70] This recognition was later developed by Lorentz who separated the polarisation of material dielectrics from that of the ether.

[71] Hertz [1962] 195.

[72] Hertz "On the Fundamental Equations of Electromagnetism for Bodies at Rest", Hertz [1962] 199 ff.

[73] On Helmholtz's polarisation: Helmholtz [187]) 615-618; Helmholtz [1881]. See also papers XXXII-XLI, in: Helmholtz [1882].

[74] Hertz, "On the Fundamental Equations of Electromagnetism for Bodies at Rest", in: Hertz [1962] 195-241 Section 4; especially "Isotropic Non Conductors", 202 ff.

[75] Helmholtz [1881] 818 ff. Hertz [1962], Introduction , 25.

[76] Hertz [1962] 25.

[77] Hertz had presented his own interpretation of Maxwell's ethereal constants beginning with his early 1884 paper and repeated it again in 1890. The dielectric constant and the magnetic constants are for him external constants, i.e., they can have only relative values with respect to the ethereal constants; accordingly, he conventionally makes to one the values of ethereal constants equal one, with the consequence that in this case electric force and polarisation coincide; another piece of evidence for the correlation between his 1884 theory and his turning point in 1887-88.

[78] Cazenobe [1980] refers to polarisation theories in his analysis of Hertz's approach to the decisive experiment. Hertz's and Helmholtz's PT are estensively examined in: Buchwald [1994] 375-388.

[79] According to accurate dating in a recent essay (Doncel [1991]), the majority of this paper was written shortly before 21 January 1888.

[80] Hertz [1962] 107. The English translation is here and in many other parts inaccurate. I present above a more accurate non-verbatim version and I quote here the original: "Wirken veränderliche elektrische Kräfte im Innern von Isolatoren, deren Dielektricitätsconstante merklich von Eins verschieden ist, so üben die jenen Kräften entsprechenden Polarisationen elektrodynamische Wirkungen aus. *Eine andere Frage aber ist es, ob auch in Luftraum veränchliche elektrische Kräfte mit Polarisationen von elektrodynamischer Wirksamkeit verknüpft sind. Man hat die Folgerung ziehen können, daß, wenn diese Fragen zu bejahen ist, die elektrodynamischen Wirkungen sich mit enldicher Geschwindigkeit ausbreiten müssen*" (Hertz [1891] 115; my Italics).

[81] Helmholtz [1870] 573, 577, 625, 627.

[82] Hertz [1962],107; paper no. 7, dated: " February 1888.

[83] Hertz, Introduction, in: Hertz [1962] 6.

[84] I remarked this omission in 1975 (D'Agostino [1975] 310, 311).

[85] Here is Hertz's original (in italics the omitted sentence): "Wären für eine bestim-

ten Isolator die erste und die zweite Voraussetzung als richtig erwiesen, so wäre gezeigt, daß sich in diesem Isolator Wellen der von Maxwell vermuteten Art forpflanzen könnten, mit einer endlichen Geschwindigkeit, welche vielleicht von der des Lichtes sehr weit abwiche. *Dies könnte aber nicht sehr überraschen, nicht mehr, als etwa der längst bekannte Umstand, daß sich im Drähten die elektrische Erregung mit grosser aber endlicher Geschwindigkeit fortpflanze.* Ich musste mir sagen, daß der Kernpunkt, der Sinn und die Besonderheit der Faraday'schen und damit der Maxwell'schen Anschauung in der dritten Voraussetzung liege,   daß es also ein würdigeres Ziel sei, wenn ich mich geradeswegs auf diese wandte" (Hertz [1891] 7; my italics).

[86] Hertz [1962] Introduction 7.

[87] Concerning this point, I disagree with Buchwald's thesis (Buchwald [1994] 262-266) on the point that Hertz awaringly changed his 1888 logic in his 1892 Introduction (See D'Agostino "Hertz's View on the Methods of Physics: Experiment and Theory Reconciled?" in: Baird, Hughes & Nordmann [1998] 89-102.

Chapter 7

[1] Hertz, *Wiedemann Annalen* 23 (1884) 84-103; Hertz [1896] 273-290.

[2] Hertz [1896] 274.

[3] D'Agostino [1975] 288 ff. The principle was extended to magnetic forces, in which case it affirms the identity of the loadstone magnetism with that induced by changing electrostatic forces, i.e., by the Maxwellian displacement current.

[4] Hertz [1896] 274.

[5] Hertz [1896] 274: "This principle is the necessary presupposition and conclusion of the chief notions which we have formed in general of electromagnetic phenomena. According to Faraday's idea, the electric field exists in space independently of and without reference to the methods of its production; whatever therefore be the cause which has produced an electric field, the actions which the field produces are always the same. On the other hand, by those physicists who favour Weber's and similar views, electrostatic and electromagnetic actions are represented as special cases of one and the same action-at-a-distance emanating from electric particles. The statement that these forces are special cases of a more general force would be without meaning if we admitted that they could differ otherwise than in direction and magnitude, that is, in the nature and mode of action."

[6] Faraday's conception is opposite to Helmholtz's conception of the correlation between matter and force: D'Agostino [1975] 292.

[7] Hertz [1896] 274.

[8] Hertz [1896] 274.

[9] Hertz [1896] 275.

[10] Hertz [1896] 276.

[11] Hertz [1896] 276.

[12] Hertz [1896] 278.

[13] Hertz [1896] 279.

[14] Hertz [1896] 281.
Hertz [1896] 285.

[15] Hertz [1896] 284.

[16] Hertz [1896] 288.

[17] Hertz [1896] 288-289.

[18] Hertz [1896] 274.

[19] Heinrich Hertz, "On the Finite Velocity of Propagation of Electromagnetic Action", *Sitzungsber. d. Berl. Akad. d. Wiss.* (2 February 1888); in *Wiedemann's Ann.*, 34 [1888], 551; in Hertz [1962] 107-123, 122.

[20] E. Aulinger, *Wiedemann's Annalen*, 31 [1887] 121.

[21] L. Boltzmann, *Wiedemann's Annalen*, 31 [1887] 598.

[22] Hertz [1896] 288: "The system of forces given by these equations is Maxwell's. Maxwell found them by considering the ether a dielectric, in which a changing polarisation produces the same effects as an electric current. *We have reached them by other premises, generally accepted even by the opponents of the Faraday-Maxwell view*"(my italics).

[23] Cazenobe [1980], argues that Hertz's 1884 derivation was a mistake because a true theory cannot be derived from a false premise. According to Cazenobe this explains Hertz's silence in 1888 as far as his 1884 derivation is concerned. In my opinion Cazenobe's argument does not adequately consider Hertz's above remarks on the logical status of his derivation Concerning its correctness, let us note that a Hertz-type derivation of Maxwell's equation is presented by Feynmann, Leighton and Sands [1965] See also a modern rendering of Hertz's derivation by Zatkis [1964].

[24] Hertz [1896] 288.

[25] The same thesis, presented in my 1975 paper, was differently evaluated by historians. Cazenobe (Cazenobe [1980] 346, 359) found it untenable, because he argued that Hertz's records indicate the contrary. He was inclined to think that the relevance of the 1884 paper is due to more than the historian's linear reconstruction. Jungnickel and McCormmach, in their outstanding work, refer to my thesis, and they also quote Planck's remarks on the subject (Jungnickel & McCormmach [1986] 47).
O'Hara and Pricha in their interesting and well documented work (O'Hara and Pricha [1890]) repeatedly refer to my evaluation of Hertz's 1884 essay and they also quote other paits of my 1975 work. Mulligan in his paper (Mulligan [1987]) agrees

with my view of Hertz [1884].

[26] Hertz [1962]107-123,122.

[27] Hertz [1962] 195-240.

[28] Hertz [1962] 236.

[29] D'Agostino [1975] 316.

[30] Hertz [1962] 210: "In the case that ether polarisations and force coincide".

[31] Hertz, Introduction, [1962] .25.

[32] Hertz [1962] 201, 200.

[33] There he defined the velocity of propagation of electric and magnetic oscillations as an "innere Constant" (Hertz [1895] 312).

[34] Hertz, "On the Fundamental Equations of Electrodynamics for Bodies at Rest", *Gottinger Nachr.* (19 March 1890); in *Wiedemann's Annalen*, 40 [1890] 577; Hertz [1962] 195-240.

[35] Hertz, "On the Fundamental Equations", in: Hertz [1962] 196-197.

[36] See"Hertz's experiments on Electromagnetic Waves", Section 5 B in this book.

[37] D'Agostino [1975] 295; Buchwald [1994] 197-198.

[38] D'Agostino [1975] 291.

[39] Hertz "On Electromagnetic Effects Produced by Electrical Disturbances in Insulators",in: Hertz [1962] 95-I06, 98,105.

[40] Hertz [1962] 288. Also, D'Agostino [1975,] 291.

[41] Hoppe [1928] 612.

[42] Hoppe [1928] 612.

[43] Buchwald [1994] 193-199.

Chapter 8

[1] Hertz [1962] 195-249; Hertz [1962] 241-268.

[2] Hertz, Introduction [1956] 1-41.

[3] The term *Bild* has a noble tradition since it was used by Immanuel Kant in his *Critique of Pure Reason*, and it was taken up with different shades of meaning by scientists such as Helmholtz, Hertz, Boltzmann, Schrödinger et al, not to mention the philosophical Kantian tradition.

[4] By this remark I do not mean that past Hertzian scholarship, although limited to particular aspects of his work, even to mere technical ones, is not a healthy and useful one. In fact it represents the necessary initial tool for fully plunging into Hertz's fundamental contributions to physics.

5 Hertz [1956] 8.
6 Hertz [1956] 12.
7 Hertz [1956] 26, 41.
8 Hertz [1956] 25.
9 Hertz [1956] 14.
10 Hertz [1956] 14.
11 Hertz [1956] 16.
12 Hertz [1956] 18.
13 Hertz, Introduction [1956] 2,3.
14 Hertz [1956] 18.
15 Hertz [1956] 22.
16 Hertz [1956] 23.
17 Hertz [1956] 23. In this amazing statement Hertz contrasts scientific men, his colleagues, to thoughtful minds, i. e., philosophers, thus producing evidence for a gap between the philosophical tradition and his contemporary science. Somebody might be surprised to discover that a leading physicist such as Hertz, who had recently given the highest contributions to nineteenth-century science, expressed such an engaging view concerning metaphysics and science.
18 Hertz [1956] 24.
19 Hertz [1956] 224, § 599.6. "Continually recurrent motions and therefore cyclical motions are frequently concealed motions, because when existing alone they cause no change in the mass distributions, nor therefore in the appearance of things" Conversely, concealed motions are almost always cyclical. Hertz, Definition 3 [1956] 226: "That part of the energy of a conservative system which arises from the motion of its visible masses is called the kinetic energy of the whole system"; Notation 606: "The energy of the concealed cyclical partial system (is) the potential energy of the whole system". And elsewhere (Hertz [1956] 277, § 607): "The kinetic and the potential energy of a conservative system do not differ in their nature; that energy which from one particular standpoint of our conception or knowledge is to be denoted as potential is from a different standpoint...denoted as kinetic".
20 Hertz [1956] 224, § 599.6 : " A system which contains no other concealed masses than those which form adiabatic cyclical systems is called a conservative system.... every conservative system may be regarded as consisting of two partial systems, of which one contains all the visible masses, the other all the concealed masses of the complete system...".
21 D'Agostino [1975] 295.
22 Hertz [1956] 25.

23 Hertz [1956] 25.

24 Catherine Chevalley (Chevalley [1991] 552-567) argues that hidden quantities have in Hertz a role comparable to virtual entities in our physical theories. This interpretation is evidently of theabove b type. In my opinion, Hertz's uncertainty on the epistemological status of hidden quantities is to be considered on the same level with Helmholtz's ambiguous interpretation of the Kantian transcendentalism, a point that Paul Hertz and Morris Schlick examined in their 1921 Centenary Edition of Helmholtz's epistemological writings (Helmholtz [1977]).

25 Hertz [1962] 21.

26 Hertz [1956] 4.

27 Hertz [1956] 4.

28 Hertz [1962] 21.

29 Hertz [1962] 21.

30 Königsberger[1965] 293. In Helmholtz's eyes, theoretical physics was also an empirical science and he struggled "to break down the barrier between experimental and theoretical physics" (Königsberger[1965] 284). Helmholtz's aversion towards a certain kind of abstractness stemmed from his polemic against the late followers of German Natürphilosophie.

31 Helmholtz, preface, in: Hertz [1956] xvi. Helmholtz never ceased to adhere to his ownconceptions of elektrodynamics. In his preface he is still unwilling renounce to his double force view (preface, VIII).

32 Chevalley [1991] 558-559, *Glossaire*.

Chapter 9

1 Boltzmann [1974] 5-12.

2 Boltzmann [1974] 10.

3 Boltzmann [1974] 11.

4 Boltzmann [1974] 77-100.

5 In 1898 Boltzmann also presented to the seventieth *Versammlung der Naturforscher* a critical comment on a special instance of Hertz's hidden-masses theory. (Boltzmann [1974] 90, note)

6 In two essays he defended atomism on epistemological grounds and in his 1897 "On the Question of Objective Existence of Processes in Inanimate Nature" (Boltzmann [1974] 57-76), he attempted a coherent presentation of an evolutionary epistemology.

7 Boltzmann [1974] 83.

8 Controversies exist about the English translation of the German word *Bild*,

which was originally used by Kant in his *Critique of Pure Reason*. Robert S. Cohen (preface, Hertz [1962] X) objects to rendering it with the English "immage", noticing that Braithwaite uses "internal picture". It is remarkable that Hertz, in describing *Bilder* as "representations", preferred to use consistently the word "Darstellung", rather than "Vorstellung" to indicate the active participation of the mind (Janik & Toulmin [1973] 140).

[9] Boltzmann [1974] 83.

[10] Boltzmann [1974] 87.

[11] Boltzmann [1974] 90-91.

[12] Boltzmann [1974] 91.

[13] Boltzmann [1974] 94.

[14] Boltzmann [1974] 94.

[15] Boltzmann [1974] 96.

[16] Fasol-Boltzmann [1990] 12-13.

[17] Boltzmann [1974] 101-128.

[18] Boltzmann [1974] 105.

[19] Boltzmann [1974] 105.

[20] Boltzmann [1974] 105

[21] Boltzmann [1974] 159-172.

[22] Boltzmann [1974] 166.

[23] Boltzmann [1974] 166.

[24] Boltzmann [1974] 169.

[25] Boltzmann [1974] 169.

[26] Boltzmann [1974] 111.

[27] Boltzmann "On the fundamental Principles", [1899], reprinted in: Boltzmann [1974] 119.

[28] Hertz, *The Principles of Mechanics presented in a new Form*, in : Hertz [1956] 26.

[29] Boltzmann [1974] 117.

[30] Boltzmann [1974] 118.

[31] Boltzmann " On a Thesis of Schopenhauer" [1905], Boltzmann [1974] 185-198, 195.

[32] Boltzmann "On the Principles of Mechanics" [1900]; Boltzmann [1974] 129-152, 133.

[33] Boltzmann [1974] 135.

[34] Boltzmann, "Fundamental Principles and Equations of Mechanics", [1899];

reprinted in: Boltzmann [1974] 108.

35 Boltzmann [1974] 68.

36 Kuhn [1978] 44.

37 Kuhn [1978] 67.

38 Kuhn [1978] 70. I believe that Boltzmann's *Bild*-conception of theory, pointing as it was to developments beyond his conception of generalised mechanics, was responsible for this prevarication.

39 Jammer [1966] 18,19.

40 Bergia [1988].

41 Bergia [1988] 227.

42 In this sense I accept Kuhn's remarks (Kuhn [1978] 70) that for Boltzmann probability calculus was primarily "a technique for evading paradoxes" and that "the mechanicistic approach to gas theory...exemplified by the H-theorem was always...[his] fundamental tool".

43 Boltzmann, "Lectures on the Principles of Mechanics" [1904]; reprinted in: Boltzmann [1974] 259.

44 Concerning Boltzmann's polemic with Planck: Meyenn [1981] 105-106.

45 J. Pouthas and J. Oms, "L. Boltzmann and the second law of the Theory of Heat", in: Sexl & Blackmore [1982].

46 Boltzmann, Vorwort [1981] IV.

Foreword to Part Three

1 Examples are: Eintein's reaction to the Michelson-Morley experiment and, in general, his attitude towards the problem of finding experimental support for his theories.

2 Jammer [1966] 109.

3 Fine [1986] 2-4.

Chapter 10

1 Bunge [1973] 204-205.

2 Fadner [1985] 836.

3 Fadner [1985] 836.

4 Fadner [1985] 831.

5 Bunge [1973] 204-205.

[6] For The QM Case: Bunge [1973].

[7] Einstein [1923] a).

[8] Einstein [July 1912] 1250-1251.
"Nach meiner Meinung muß Man an der Postulate 3 unbendingt fest gehalten werden, solange nicht zwingende Gründe dagegen vorliegen; sobald wir von diesem Postulat abgeben, wird die Mannigfaltigkeit der Möglichkeiten unübersehbar"

[9] Einstein, [February 1912].

[10] Pais [1982] Chap. 9·

[11] The 1907 EP expressed a limited covariance of the gravitation field; as in the special theory, a magnetic field could be switched on by a linear coordinates, transformation, so it was for gravitation in GR.

[12] EP states the equivalence between a gravitational and an acceleration field. In its local form, EP limits the equivalence to infinitesimal spatial regions.

[13] *Zurich Notebook* [ea 3006]; a copy was kindly forwarded to me by the Einstein Archive in Princeton.

[14] Einstein [March 1912] 456. Idem, p.1251.
"Schritt entschließe Ich mich deshalb schwer, weil Ich mit ihm den Boden des unbedingten Aequivalenzprinzips verlasse. Es scheint, daß sich letzteres nur für unendlich kleine Felder aufrecht erhalten läßt".

[15] Norton [1985].

[16] Norton [1985] 203.

[17] For a detailed analysis of the problem of general covariance in Einstein. See: Norton [1984].

[18] Einstein-Grossmann [1913] 233. "Es muß aber hervorgehoben werden, daß es sich als unmöglich erweist, unter dieser Voraussetzung einen Differentialausdruck gmn zu finden, der eine Verallgemeinerung von df ist, und sich beliebigen Trasformationen gegenüber als Tensor erweist".

[19] Einstein [1923] b).

[20] Einstein-Grossmann [1913].

[21] Einstein [1916].

[22] Eistein treated the same problem using a similar procedure in: Einstein [1916] 777-778.

[23] There are no rigid rods in an extended gravitational field.

[24] Einstein [1936] d), 20.

[25] Einstein [1936] 21.

[26] Einstein [1936] 21.

[27] Einstein [1936] 18. "These notions and relations, although free statements of our thought, appear to us stronger and more unalterable than the individual sense expe-

rience itself".

Chapter 11

[1] Remarkably, recent historical-epistemological studies on the foundation problems of quantum physics have adopted this viewpoint (Cattaneo & Rossi, [1991]).

[2] This conceptual ambiguity can be interpreted as an example of "concepts in flux" (Elkana [1970]).

[3] Bridgman [1959] 335 ff.

[4] Einstein [1949] b) 679.

[5] Jammer [1966] 198. Quoted from the *Archive for History of Quantum Physics*, [1963].

[6] Einstein [1923] 482.

[7] Einstein [1923] 483.

[8] Born [1909].

[9] Maltese & Orlando [1994].

[10] Einstein [1923] 485, 489.

[11] Einstein [1923] 487.

[12] Einstein [1934] 307-08.

[13] Einstein [1934] 307.

[14] Einstein. [1950] 63-65.

[15] D'Agostino & Orlando [1994].

[16] Einstein [1950] 64.

[17] Einstein [1950] 81.

[18] D'Agostino & Orlando [1994])

[19] Einstein [1923] 484.

[20] Levi-Civita [1917].

[21] Janssen [1988] 351.

[22] Einstein [1918] 480; Vizkin [1986] 303.

[23] Einstein [1923] 483.

[24] Einstein [1950] 94.

[25] Bergia [1991].

[26] Einstein [1949] b.

[27] Einstein [1949] b, 677 ff.

NOTES

[28] Einstein [1949] b, 678.

[29] Einstein [1949] b, 680.

[30] Einstein [1959] c, 59.

[31] In my view, a four dimensional non-affine theory does not introduce electromagnetic fields.

[32] Einstein [1959] b, 685.

[33] Einstein [1959] c, 59.

[34] Fine [1986]; Howard [1989]; Howard [1990].

[35] "It may be taken as a peculiar aspect of the often winding paths of the history of physics that Bohr actually transerred the symbolic character of his QM to the Einsteinian theory of GR, in as much as the latter made use of a not directly visualisable symbolism" (Chevalley [1993]). More than a further misunderstanding, I think that Bohr's assignment may be properly understood as his realisation that Einstein's ideal *Anschaulichkeit* had actually failed.

[36] Holton [1973] 246.

[37] Planck [1931] 15-17; Holton [1973] 244-245.

[38] Einstein, Introduction to Planck [1931] b; Holton [1973] 244.

[39] Vizgin [1986] 310.

[40] Vizgin [1986] 309.

[41] Given this situation, it is understandable that Einstein considered the meaning problem as central. Its solution would have helped to provide the geometrised UFT with a more acceptable physical basis (Vizgin [1987] 39).

[42] Vizgin [1986] 309.

[43] Chevalley [1989] a.

[44] Chevalley [1989] b; Chevalley [1993] .

[45] Consistent with the above thesis, I want to argue that Einstein's demand that completenessrequires connection between the concepts of the theory and perceptive experience can be understood as a request for a Kantian synthesis. Moreover, the failure of Einstein's method to achieve the SM requirement in its various forms represented a failed attempt at Kantian synthesis. This observation would clarify Einstein's adherence to Kantianism.

[46] Miller [1996] 215.

[47] Miller [1996] 46.

[48] Chevalley [1989] b), 151.

Chapter 12

[1] Popper [1967], Park & Margenau [1968], Park [1968]. A revaluation of Bohr's philosophy is presented in: Freundlich [1968]; Honner [1982].

[2] "In 1927 Niels Bohr, one of the greatest thinkers of the field of atomic physics, introduced the so-called principle of complementarity into atomic physics, which amounted to a 'renunciation' of the attempt to interpret atomic theory as a description of anything. I do not believe that physicists would have accepted such an ad hoc principle had they understood that is was ad hoc, or that it was a philosophical principlepart of Bellarmino's and Berkeley's instrumentalist philosophy of physics. But they remembered Bohr's earlier and extremely fruitful 'principle of correspondence' and hoped (in vain) for similar results" (Popper [1963] 100-101). I have quoted this more recent Bohr's statement for its clarity in distinguishing the role of the two principles. However, Popper's criticism of Heisenberg's indeterminacy relations (IR) occurs as early as 1934 in:Popper, New appendices [1959] revised ed. [1968], 454. See also: Popper [1982].

[3] Jammer [1974] 60.

[4] Jammer [1974] 98.

[5] Léon Rosenfeld attested that Bohr's "turn of mind was essentially dialectical, rather then reflective". One of Bohr's dictums was: "Every sentence I say must be understood not as an affirmation, but as a question". Rosenfeld [1967] 63. Holton [1973]. As a psychological remark I like to quote Niels Bohr's brother's sympathetic and amusing complaint that Niels seldom completed in his more provocative talks what he was about to say.

[6] Rudinger [1976-1986], henceforth NBCW [1976-1986].

[7] Krajewski [1977]. In 1922 Bohr himself affirmed that in his 1913 papers there are the first germs of the Cr Principle(Bohr [1922]. On Bohr's interpretations of Cr between 1918 and 1925, see: Petruccioli[1988].

[8] Jammer [1974] 111 note 94, 110.

[9] Born & Pauli, as reported by Jammer [1974] 116.

[10] NBCW [1984] 273-280; NBCW [1976-1986] Also; Bohr [1935] 845-852.

[11] "While Einstein's theory of heat radiation gives support to the postulates, it accentuates the formal nature of the frequency condition" (Bohr NBCW [1984] 275).

[12] The contradiction is inherent in the definition of the photon energy $h\nu$. It was first presented in: Bohr, Kramers, Slater, [1924] 787, and then stated by Bohr in: Bohr [1925].

[13] The expression Bohr uses is: "visualisation of the stationary states by mechanical pictures". In the paragraph "The Correspondence Principle" he illustrates his meaning of "visualisation": "Nevertheless it has been possible to construct mechanical pictures of the stationary states which rest on the concept of the nuclear atom and have been essential in interpreting the specific properties of the elements".

(Stolsenburg [1984] 276). Notice that, at this stage, Bohr uses "visualization" to mean the forminng of "mechanical pictures".

14 According to Max Jammer (Jammer [1984] 109), the assumption that QT at least formally contains CT as a limiting case, was first stated by Planck in 1906 in his well-known statement that when h tends to zero, his radiation formula becomes the classic Rayleigh-Jeans formula. However, this statement was contradicted by Bohr in 1913; according to him QT approaches CT when the differences in the quantum numbers are small compared with n. Bohr called this "a formal analogy between QT and CT". In 1918 Bohr generalised the rule for conditionally periodic systems.

15 NBCW [1984] 276.

16 NBCW [1984] 276, 277.

17 NBCW [1984] 277.

18 NBCW [1984] 277. As a matter of fact, in Bohr's essay, the list of the difficulties precedes in part that of the successes of classical theories. The logic of my presentation has suggested the reversal; however, this change does not introduce any element of unfaithfulness to Bohr's ideas.

19 NBCW [1984] 274. In the addendum, Bohr says that in the phenomena of atomic colisions, which are "incompatible with the properties of mechanical models ,[...] we must in fact even be prepared to find behaviour that is as alien to the application of the ordinary space-time pictures as the coupling of individual processes in distant atoms is to a wave description of optical phenomena"(Bohr [1925]; NBCW reprinted in:[1984] 205). On the role played by the Kramers-Heisenberg dispersion theories, and the Bohr-Kramers-Slater story of the attempt at introducing virtual oscillators, see: Jammer [1966] 345, 346 ff.

20 (Bohr to Pauli 25 November 1925, reprinted in: NBCW [1984] 224). The failure of the Bohr-Kramers- Slater theory of virtual oscllators is, of course, a turning point in the development of the meaning of Cr. A useful factual account of the Bohr-Kramers-Slater theory and the so-called Kramers-Heisenberg dispersion formula is presented by Tagliaferri [1985] 334 ff..

21 NBCW [1984] 277.

22 NBCW [1984] 276, ff..

23 NBCW [1984] 280. The long quotation is justified by the meaning I attribute to the restriction as a first step, which will be generalised in the limitation of Cm. See: Holton [1973] 130.

24 Heisenberg [1925].

25 However it should be remarked that, as a matter of fact Heisenberg's mathematical scheme for constructing the new QT was not limited to frequencies and amplitudes, but it included the Heisenberg-Born derivation rule as well as the matrix substitution for spacial coordinates.

[26] NBCW [1984] 280.

[27] NBCW [1984] 276.

[28] I have found no explicit mention in the consulted literatureof my thesis that the new meaning of Bohr's Cr in 1925 was influenced by Heisenberg's 1925 paper.

[29] Kalchar, Introduction, in: NBCW [1985] 21.

[30] NBCW [1985]; Bohr [1926-1932] 91-99. Also: *Archive for History of Quantum Physics* (AHQP), microfilm, no 36.

[31] Heisenberg [1927].

[32] NBCW [1985] 91.

[33] The "indispensability" had however been proposed as early as 1922: "Every description of natural processes *must be based* on ideas which have been introduced and defined by the classical theory" (my underlining). N. Bohr, "On the Quantum Theory of Line Spectra", *Zeitung für PhysIK,* 9, 1922, 1; my italics).

[34] Concerning the interpretation of Bohr's conception of theory as a symbolic expression, consult: Chevalley [1993].

[35] Bohr, "Atomic Theory and mechanics", in: NBCW [1984] 276 ff.

[36] These ideas are presented in: Bohr [1958]. See also: Honner [1982].

[37] Kalckar, Introduction, NBCW [1985] 26-27. Notice that in this passage Cm concerns the antinomy [superposition principle/conservation principle]. In this form it remains closer to the wave-particle antinomy of electrodynamics and to the Boh-Kramers-Slater interpretation of optical dispersion.

[38] NBCW [1985] 92.

[39] Jammer [1974] 93, 65.

[40] See: *Archive for History of Quantum Physics* [1963], interview with Heisenberg, 25 February 1963.

[41] "These reciprocal uncertainty relations were given in a recent paper by Heisenberg [*Zeit. für Phys.*, 49, 1927, 172-198] as the expression of the statistical element which, due to the feature of *discontinuity* implied in the quantum postulate, characterises any interpretation of observation by means of classical concepts" (NBCW [1985] 93, my italics). Bohr refers to discontinuity (the quantum finiteness) as something characterising Heisenberg's approach to IR, not his (Bohr's) own approach, which, I have shown, is based on the link which the quantum of action establishes between ondulatory and particulate features. The point of the difference is taken up again by Bohr some lines later, when he discusses the "gradual spreading of the wave fields" (NBCW [1985] 94), and "the statistical character of the quantum theoretical description" (NBCW [1985] 98).

[42] Bohr, "The Quantum Postulate and the recent Development of Atomic Theory", *Nature,* Supplement [14 April 1928], 580-590; reprinted in NBCW [1985] 147

43 NBCW [1985] 151. For Bohr's and Heisenberg's views of the X-ray microscope, see the interesting paper in: Kalckar[1985] 16 ff.

44 Bohr, "The Quantum Postulaten", NBCW [1985] 151.

45 NBCW [1985].

46 Bohr's concern about his Cm also involved the visualisation problem (an aspect of his problem with ST description). This aspect has been rightly emphasised by Hendry [1984], by Miller & Hendry [1984], and by Miller [1984].

47 According to Jammer, this type of criticism of Heisenberg's interpretation seems to have been raised explicitly for the first time by Ch.R. von Liechtenstern in 1954. However, I think that it amounts to the same type of argument presented by Popper in 1934 (Jammer [1974] 345).

48 Jammer [1974] 73.

49 "There seems to me a fairly obvious connection between Bohr's 'principle of complementarity' and this metaphysical view of an unknowable reality [i.e. Heisenberg's interpretation of the uncertainty Relations] a view that suggests the 'renunciation' (to use a favourite term of Bohr's) of our aspiration to knowledge, and the restriction of our physical studies to appearances and their interrelations"; (Popper, "Quantum Mechanics without the observer"in: Bunge [1967] 7-44; Also: Popper [1968] 454).

50 Popper [1967].

51 Once this point has been made I think that we should admit that Bohr's formulation of Cm is not completely unrelated to Heisenberg's interpretation of IR. Although the last interpretation is not required for the strict demonstration of Bohr's IR, the theme of the finiteness of the quantum (i.e., Heisenberg's affirmation that discreteness is the primary cause of disturbance in the otherwise unperturbed definition of position and momentum) appears here and there in Bohr's 1927 papers.

52 Hans Reichenbach was in 1929 one of the first to stress this genuine aspect of Bohr's ontology. He emphasised that the source for Bohr's Cm is primarily the structure of the theory presented by QT and not the interpretation of IR as disturbance to an otherwise undisturbed ST description. (H. Reichenbach, "Wiele und Wege der physikalischen Erkenntniss", *Handbuch der Physik*, Vol. 4, ed. H. Seiger and K. Schee, Springer, Berlin 1929, 78; reported by Jammer, [1974] 160). This view is also more generally confirmed by such an acute epistemologist as Eino Kaila (Kaila [1950]). The fact that quantisation can be explained without referring to quantum was shown, of course, by Schrodinger's wave-theory (see my following chapter).

53 Jammer [1966] 117. The "inclusion" view, Cr(1), was not, at any rate that of Bohr, at least, after 1924, point proved by Bohr's 1924 statement quoted by Jammer Jammer [1966] 118). Perhaps Cr(1) had this function in the early stages of the

"logic of discovery" of QT.
[54]Jammer [1966] 88.
[55]Tagliaferri [1985] 240.

Chapter 13

[1] Schrödinger,"Quantisierung als Eigenwertproblem", *Annalen der Physik, 79;* Schrödinger [1926] a.

[2] Schrödinger [1926] a; *Vierte mitteilung.*

[3] Schrödinger [1926] a, *Erste Mitteilung*, Sect 3.

[4] Born [1926].

[5] Schrödinger [1926] b.

[6] Schrödinger [1929] 15-16.

[7] Schrödinger [1929] 15.

[8]. Schrödinger [1932]; Schrödinger [1987] 20-36.

[9] Schrödinger [1935].

[10] Fine [1986] 64-85.

[11] According to Ruger: "Affine theory in the Forties and QM in the Thirties and Forties formed a complex of problems out of which grew, in the 1940's, Schrödinger's affine field theory and his 'late' interpretation of QM" (Rüger [1988] 378).

[12] The method of second-quantisation was developped in 1927 by P.A.M. Dirac for particles obeying Bose statistics, and later in 1928 extended to Fermi particles by E.Wigner and P. Jordan .

[13] Schrödinger [1928]; Schrödinger [1984], Vol. 3, 185-206, 185.

[14] Schrödinger [1984], Vol. 3, 205.

[15] C.N.Yang emphasises Schrödinger's reluctance at that time to use complex numbers in the role of physical quantities: Yang, "Square roots of Minus One, Complex Phases and E.Schrödinger" in: Kilmister [1987] 53-64, 54.

[16] Schrödinger, "What is an elementary particle" in Schrödinger [1984] 456-463, 459.

[17] Schrödinger [1984] 459.

[18] Schrödinger [1984] 459.

[19] Schrödinger [1953] 16.

[20] Schrödinger [1953] 24.

[21] Schrödinger [1953] 24. Schrödinger quotes his 1950 paper in *Endeavour* and his 1951 paper in the Austrian Journal *Die Pyramide*.

[22] Schrödinger [1953] 24.

[23] Schrödinger [1953] 28.

[24] Appendix 1 in this section.

[25] Appendix 2, in this section.

[26] Schrödinger [1984] 295-297, 296.

[27] Fine mantains (Fine [1986]) that Schrödinger addressed the same critique to Einstein's statistical interpretation of QM.

[28] Schrödinger [1932] 8; Schrödinger [1987] 31.

[29] Fine [1986] 81-85.

[30] Schrödinger [1951].

[31] Schrödinger [1951] 54.

[32] Schrödinger [1951] 55.

[33] Schrödinger [1951] 55.

[34] Schrödinger [1951] 56.

[35] Schrödinger [1951] 54.

[36] Schrödinger [1951] 19.

[37] Schrödinger [1951] 57.

[38] Schrödinger [1951] 17, I8.

[39] On many occasions in his writings, Schrödinger refers to his teacher Franz Exner as the first to advance doubts on energy conservation at the atomic scale and, more generally, proposed a non-deterministic view of the microworld, and a rejection of statistics in the classical sense. An early reference to Exner is in: Schrödinger [1984] Vol. 4, 295-297, 296-297.

[40] Schrödinger [1951] 21.

[41] Schrödinger [1951] 47, 48.

[42] Schrödinger [1951] 48.

[43] "I will not say that Kant's idea [transcendentalism of space and time as forms of "mental intuition"(*Anshauung*)]was completely wrong, but it was
certainly too rigid and needed modification when new possibility came to
light.This new possibility is offered by Einstein's restricted theory of relativity"; in: Schrödinger [1951] 48.

[44] Schrödinger [1951] 65.

[45] Schrödinger [1951] 53.

[46] Schrödinger [1951] 63.

[47] Schrödinger [1951] 1-2.

[48] Schrödinger, "The meaning of the Wave Mechanics" in: Einstein, De Broglie et

al.[1953] 16.

[49] Schrödinger [1951] 28.

[50] Schrödinger [1951] 28-29.

[51] Schrödinger [1951] 29.

[52] Schrödinger [1951] 40.

[53] Schrödinger [1951] 32-33.

[54] Schrödinger [1951] 40.

[55] Schrödinger [1951] 41.

[56] Schrödinger [1951] 29.

[57] Schrödinger [1951] 40. In "Are there Quantum Jumps ?", Schrödinger criticised Bohr's theory on the point that, in this theory, atomic radiative transitions occur instantaneously; Schrödinger "Are there Quantum Jumps?" in: Schrödinger [1984] 478-502, 483 .

[58] Schrödinger [1951] 39.

[59] One could here refer to one of Heisenberg's numerous statements against Schrödinger's positions; e.g., on the electron transition between two stationary levels in an atom: "Schrödinger therefore rightly emphasises that... such processes can be conceived of as being more continuous then in the usual picture, but such an interpretation cannot remove the element of discontinuity that is found everywhere in atomic physics: any scintillation screen or Geiger counter demonstrates this element at once. In the usual interpretation of quantum mechanics is contained the transition from the possible to the actual. Schrödinger himself makes no counterproposal as to how he intends to introduce this element of discontinuity, everywhere observable, in a different manner from the usual interpretation" (Heisenberg [1958] 84). Evidently, Heisenberg did not agreewith Schrödinger's proposal of two levels of languags, and with his idea that an element of discontinuity is acceptable at the observational level.

[60] Schrödinger [1951] 41. The highly generalised type of field equations are the wave functionals resulting from second quantisation.

[61] Schrödinger [1951] 19.

[62] Schrödinger [1958] 169.

[63] Schrödinger [1958] 169.

[64] "It is true that in thinking about the atom, in drafting theories to meet the observed facts, we do very often draw geometrical pictures on the black-board, or on a piece of paper, or more often just only in our mind, the details of the picture being given by a mathematical formula with much greater precision...but the geometrical shapes....are not anything that could be directly observed in the real atoms. The pictures are only a mental help, a tool of thought..." ; (Schrödinger [1951] 21).

[65] Schrödinger [1951] 40.

66 Schrödinger [1951] 26. The passage is quoted with the author's comment in: Arendt [1958]; Arendt [1988] 213. In Arendt's view: "the new universe of science presented by Schrödinger is not only practically inaccessible but unthinkable as well". As we have seen, Schrödinger in effect distinguishes between the capacity of thinking a perfect model from the conviction that the model represent the true universe of science. A perfect model is thinkable, while the latter is not only unthinkable but even beyond such ideas as Kant's *Noumenon*.. For Schrödinger, the metaphorical example of a winged lion symbolises a partially thinkable objec.

67 Schrödinger [1951] 25.

68 Schrödinger might refer to the following passage in Boltzmann's "On the significance of theories: "I am of the opinion that the task of theory consists in constructing a picture of the external world that exists purely internally and must be our guiding star in all thought and experiment; that is *in completing, as it were,the thinking process* and carrying out globally what on a small scale occurs within us whenever we form an idea"(Boltzmann [1974] 33).

69 Schrödinger [1951] 25.

70 On such matters, Schrödinger preferred the positivist's fidelity to phenomena, ignoring the quest for knowledge which is not attainable in principle: "I fully agree [with the positivist] that the uncertainty relation has nothing to do with incomplete knowledge. It does reduce the amount of information attainable about a particle as compared with views held previously. The conclusion is that these views were wrong and that we must give them up"; (Schrödinger [1950] 456).

71 For an interesting thesis on Hertz's Bild, see: Chevalley [1991] 549 ff.

72 Hertz: "Die Bilder von welche wir reden, sind unsere Vorstellungen von den Dingen; sie haben mit den Dingen die eine wesentlich Übereinstimmung, welche in der Erfüllung der genanten Forderung liegt, aber es ist für ihren Zweck nicht nötig, daß sie irgend eine weitere Übereinstimmung mit den dingen haben" (Hertz[1895], *Einleitung*, 1-2). Let us compare this passage with Schrödinger's statement in 1928:"Ich glaube dazu kann man doch einige erkenntnistheoretische Trostworte sagen. Wir dürfen nicht vergessen, daß die Bilder und Modelle schließlich doch keinen anderen Zweck haben, als alle prinzipiell möglichen Beobachtungen an ihnen aufzuhangen"(Schrödinger [1984] 294).

73 Schrödinger [1951] 24.

74 Schrödinger [1958].

75 Schrödinger [1927].

76 I am grateful to Helge Kragg for this objection .

77 Some of these themes, not all of them in line with his predominant interest of the moment, Schrödinger just touched on and left unexplored. One example is offered by a letter from F. London to Schrödinger, quoted by C. N.Yang in his "Square root of -1, Complex Phases and Schrödinger"; in: Kilmister [1987] 53-63, 62.

[78] Schrödinger [1953] 28.

[79] J. Dorling inclines to consider the physicists' dismissal of Schrödinger's views as "largely a sociological accident"; in: Dorling [1987] 39.

[80] See among others, the papers by J. Dorling, J. S. Bell and C. N. Jungin; in: Kilmister [1987]. See also: A.O. Barut, *Ann.der Physik*, (Schrödinger issue [1987]) and *Foundation of Physics*, (Schrödinger issue [1987]).

[81] Schrödinger [1984] 461.

[82] Schrödinger [1935] 65.

[83] Landau & Lifshitz [1958] 221.

Chapter 14

[1] Einstein [1949].

[2] Schrödinger [1951] 47.

[3] Schrödinger [1951] 24, 25.. Very probably he refers to Hertz's and Boltzmann's *Bild*-conception.

[4] Schrödinger uses this word in his German articles and often quotes his teacher Franz Exner (Boltzmann's student) and even Boltzmann himself.

[5] Boltzmann [1974] 96.

[6] Chevalley [1985] 265.

[7] Schrödinger [1951] 8.

[8] I refer mainly to Einstein's completeness, as stated for instance in: Einstein [1949] 665-688.

[9] A .Fine [1986] 80.

[10] Chevalley [1985] 251-292.

BIBLIOGRAPHY
PRIMARY SOURCES

AHQP:*Archive for History of Quantum Physics* ,
American Philosophical Society Library, Philadelphia,
Accademia dei Quaranta, Roma, and elsewhere.
1963 "Interview with Einstein", February 15.

Ampére A. M.
1826: *Théorie des Phénomènes Electro-dynamiques uniquement déduite de l'Expérience*, Paris.
1921: "Sur la détermination de la formule qui représente l'action mutuelle de deux portions infinement petites de conducteurs voltaiques", in:*Mémoires sur l'Electromagnetisme et l'Electrodynamique*, Gauthiers - Villars, Paris.

Boltzmann L.
1891: *Vorlesungen über die Maxwellsche Theorie der Elektricität und des Lichtes*, I. Theil, Leipzig.
1905: *Populäre Schriften (1892-1905)*, Leipzig.
1974 :*Theoretical Physics and Philosophical Problems* (Mc Guinness Edit.) [A collection of Boltzmann's Essays from: *Populäre Schriften* (1892-1905), and from: *Vorlesungen über die Principe der Mechanik* , (1897-1904)], Reidel Pbl.Co.

Bohr N.
1925: "Atomic Theory and Mechanics", *Nature* (Suppl), II 6 (1925) 845-852; in: *Niels Bohr Collected Works [1984]* 273-280.
1927: "The Quantum Postulate and the Recent Development of Atomic Theory", in: *Niels Bohr Collected Works [1985]* 91- 99. Also in: AHQP, n. 36.
1958: *Atomic Physics and Human Knowledge*, Wiley.

Bohr N, N.A. Kramers, and J.C. Slater.
    1924 "The Quantum Theory of Radiation", *Philosophical Magazine* 37 (1924) 785-802.

Born M.
    1909: "Die Theorie der stärren Elektrons in der Kinematik des Relativitätsprinzip", in: *Annalen der Physik,* 30 (1909) 1-56.
    1926: "Zur Quantisierung der Stossvorgange", *Zeitung für Physik,* 37 (1926) 711-726

Bridgman P.W.
    1959: "Einstein's Theory and the Operational Point of View", in: Schilpp A.(ed.) *Albert Einstein, Philosopher Scientist*, Harper & Brothers Pub.

Clausius R.
    1882: "Über die verschiedenen Maßysteme...", *Annalen der Physik und Chemie*, N.F., Band XVI.

Clifford W. K.
    1878: *Elements of Dynamics*, Part 1, Kinematic, London.

Fourier J.
    1822 :*Theorie analytique de la chaleur*, Paris .

Einstein A.
    1912 : February , "Lichtgeschwindigkeit und Statik des Gravitationsfeldes." *Annalen der Physik,* 38 (1912) 355-369.
    1912:March,"Zür Theorie des statischen Gravitationsfeldes." *Annalen der Phys*ik, 38 (1912) 443-458.
    1912: July ,"Relativität und Gravitation. Erwiderung auf eine Bemerkung von M. Abraham." *Annalen der Phys*ik, 38 (1912) 1059- 1064.
    1917: Über die spezielle und allgemeine Relativitätstheorie (gemeinverständlich), Braunschweigs.

1918: "Dialog über Einwände gegen die Relativitätstheorie", *Die Natürwissenschaften.* , vo.6, pp. 697-702. Quoted in: Norton [1985].

1923:a)"Grundgedanken und Probleme der Relativitäts theorie", Stockholm, Imprimerie Royal,

1923: b) "Fundamental Ideas and Problems of the Theory of Relativity", lecture delivered to the Nordic Assembly of Naturalists at Gothenburg, 11 July, 1923, in: *Nobel Lectures.*, Physics 1901-1921, Elsevier 1967.

1933 : "On the Method of Theoretical Physics". The Herbert Spencer Lecture delivered at Oxford, 10 June, 1933. Oxford Clarendon Press.

1934: "On the Method of Theoretical Physics". Repr. in: *Philosophy of Science*, vol.1, 162-169.

The German Text is in *Mein Weltbild* , Amsterdam, Querido 1934 pp 176-187.A transl. :*The World as I see It*, NY, Cocivi-Friede, 1934. The Phil Library, NY 1949, 269-290 .

1936 a): "Physik und Realität", *Zeitschrift für freie deutsche Forschung*, Paris vol. 1(1936) n 1, 2.

1936 b): " Physics and Reality "; a transl,, *Franklin Ist. Journal*, vol 221, N 3 (1936) 313-347.

1936 d): *Physics and Reality*, Essays in Physics Philosophical Library.

1949 a): "Reply to Criticism", in: Schilpp (Edit) *Albert Einstein Philosopher Scientist* , Lybrary of Living Philosophers.

1950: " Physics and Reality", Repr.inted in: *Out of my Later Years* , Philosophicaal Library, NY ; 59-97.

1957: "Fisica e Realtà" , in : *Idee e Opinioni*, Schwarz, Milano; Italian translation of :Einstein [1950].

1959: *Albert Einstein Philosopher Scientist*, Schilpp P. Arthur (Edit.) Harper Torch Books, 2 Vols.

1959 b): "Reply to Criticism", in: Schilpp (Edit), *Albert Einstein Philosopher Scientist*, Harper & Brothers Publ. NY, 2 vols, vol 2, 665-688.

1959 c): "Autobiographical Notes,in: *A.Einstein Philosopher Scientist*, Harper & Brothers Publishers,NY 1959, 2 vols;

vol.1., 1-95.

1967:*Relatività: esposizione divulgativa*, Boringhieri, Torino, (Italian translation of: "Über die spezielle und allgemeine Relativitätstheorie (gemeinverständlich)", Einstein [1916]).

1988: "Idee e Problemi fondamentali della Teoria della Relatività" (ital. transl.of Einstein [1923] a), in: Albert Einstein. *Opere scelte*, a cura di Enrico Bellone. Bollati Boringhieri, Torino..

Einstein A.-Grossmann M.,
1913: "Entwurf einer verallgemeinerten Relativitätstheorie und einer Theorie der Gravitation." *Zeit.schrift Mathematische Physik.* 62 (1913) 255.

Einstein, De Broglie et al
1953: *Louis DeBroglie, Physicien et Penseur*, Albin Michel.

Fitzgerald G.F.
1902: *The Scientific Writings*, Dublin and London.

Gauss C.F.
1832: "Intensitas vis magneticae ad mensuram absolutam revocata", *Göttingschen geleherte Anzeigen,* 24 December.
1867: *Werke,* Göttingen.

Gauss C.F., Weber E.
1840: "Messung starker galvanische Ströme bei geringe Widerstande nach Absolute Maass ", *Beobakten der Magnetische Vereins,* Jahre 1893. Reprinted in:
1893: *Wilhelm Weber's Werke* [1893].

Giorgi G.,
1937: "Unità (sistemi di)", *Enciclopedia Italiana,* Rizzoli 36 vols., vol. 34.

Heisenberg Werner
   1927: "Uber den anschulichen Inhalt der quantentheoretischen Kinematik und Mechanik", *Zeitschrift fur Phys.ik*, 49 (1927)172-198.
   1958: *Physics and Philosophy*, Allen & Unwin.

Helmholtz H.
   1869: "Über die Dauer und den Verlauf der durch Stromschwingungen induzierten elektrischen Strome", reprinted in: Helmholtz [1882] 429-462.
   1870: "Über die Theorie der Elektrodynamik. Erste Abhandlung. Über die Bewegungsgleichungen der Elektricität für ruhende leitende Körper", *Borchardt's Journal fur die reine und angewandte Mathematik*, 72 (1870) 57-129; reprinted in Helmholtz [1882] 545-628.
   1878: "Die Tatsachen in der Wahrnehmung" in: Helmholtz, *Populären wissenschaftliche Vorträge*, 2 vol, 1865 and 1871.
   1977: "The Facts in Perception", English Translation of Helmoltz [1878], in: Cohen & Helkana [1977] 115-185; reprinted in Helmholtz [1977] 115-185.
   1881: "Über die auf das Innere magnetisch oder diëlektrisch polarisirter Körper wirkenden Kräfte", in : Helmholtz [1882] 798-822.
   1882: *Wissenschaftliche Abhandlungen*, Erster Band, Leiptzig.
   1865 and 1871: *Populäre wissenschaftliche Vorträge*, 2 vols.
   1865 and 1871: *Vorträge und Reden*, vol. 2, 215-247, 387-406.
   1977: *Epistemological Writings*, The Paul Hertz/Moritz Schlick Centenary Edition of 1921, with Notes and Commentary by the Editors, newly translated by M.F. Lowe. Edited, with an Introduction and Bibliography, by R.S. Cohen and Y. Elkana;*Boston Studies in the Philosophy of Science*, vol. XXXVII, Reidel.

Hertz H.
>1891: *Untersuchungen über die Ausbreitung der elektrischen Kraft*, First edition.
>1894: *Prinzipien der Mechanik in neuen Zusammenhang dargestellt* , Leipzig.
>1895: *Gesammelte Werke*, Band I, *Schriften Vermischten Inhalts*, Leipzig.
>1896: *Miscellaneous Papers* , London.
>1956: *The Principles of Mechanics Presented in a New Form* ; authorized English translation by D.E. Jones and T.E Valley, 1879, Dover reprint.
>1962: *Electric Waves, being Researches on the Propagation of Electric Action with Finite Velocity through Space* ; authorised English translation by D.E. Jones; Dover Reprint New York.

Kelvin (Lord), Thomson W.
>1981: *Popular Lectures and Addresses*, vol. 1, London

Kirchhoff G.R.
>1857: "Über die Bewegung der Elektricität in Drähten,",
>1882: *Gesammelte Abhandlungen* , Barth, Leipzig.

Larmor J. (ed.),
>1907: *Memoir and Scientific Correspondence of the Late Sir George Gabriel Stokes*, 2 vols., Cambridge 1907, vol. 2.
>1937: *The origins of Clerk Maxwell's Electrical Ideas, as Described in Familiar Letters to William Thomson*, Cambridge University Press.

Levi-Civita T.
>1917: "Nozione di Parallelismo in una varietà qualunque", *Rendiconti del Circolo Matematico di Palermo* , 42 (1917) 137-205.

Lodge Oliver
> 1894: *The Work of Hertz and Some of His Successors*, London.

Mach Ernst,
> 1926 *The Principles of Physical Optics ; an Historical and Philosophical Treatment* , English Translation.
> 1953: *The Principles of Physical Optics; an Historical and Philosophical Treatment* , New Dover edition.
> 1894: *Popular Scientific Lectures*, English translation.
> 1943:*Polular Scientific Lectures*, English translation; Reprinted: The Open Court Pubishing Co.

Maxwell J.C.
> 1850: "On the Equilibrium of Elastic Solids", in: Maxwell [1954] vol 1, 31-73.
> 1855: "On Faraday's Lines of Force", in: Maxwell [1954] vol 1, 155-229.
> 1861: "On Physical Lines of Force", *Philosophical Magazine*, 21; Part I, March 1861: Part II, April and May 1861.
> 1862: "On Physical Lines of Force", *Philosophical Magazine* January and February 1862: Part III ; also in Maxwell [1954] vol. I, 451-513.
> 1864: "A Dynamical Theory of the Electromagnetic Field", *Transaction Royal Society* , CLV, received: October 27, read: December 8, 1864; also in: Maxwell [1954] vol. I.
> 1868: "On a Method of Making a Direct Comparison of Electrostatic with Electromagnetic Force; with a Note on the Electromagnetic Theory of Light", *Philosophical Transactions,* Royal Society of London, 158(1868) 643-658; also in: Maxwell [1954] vol. 2, 125-143.
> 1890: *The Scientific Papers of James Clark Maxwell,* W.D. Niven (ed.), Third Edition, 2 vols.
> 1954: *The Scientific Papers of James Clark Maxwell,* Dover Reprint; 2 Vols.
> 1954: *A Treatise on Electricity and Magnetism,* Dover reprint

2 Vol.
1877: *Matter and Motion.*
1920(circa): *Matter and Motion*, Reprinted, with Notes and an Appendix by Sir Joseph Larmor; (undated) Dover reprint.
1995: *Scientific Papers and Letters*, Vol. 2: 1862-1873, Cambridge University Press.

Maxwell I.C. and Fleming Jenkin
1863: "On the Elementary Relations between Electric Measurements", Appendix C, in:*The Report of the British Association for the Advancement of Science*, 1st series, 32, 111-176.
1871(circa): "On the Mathematical Classification of Physical Quantities", (undated) Maxwell [1954] Vol. 2, 257-266.

Planck M.
1931 a): "J.C.Maxwell in seiner Bedeutung für die theoretische Physik in Deutschland", *Naturwissenschaften*, 19 (1931) 889-894; reprint in:*Physikalische Abhandlungen und Vorträge,,* 3 (1931) 352-57.
1931 b): "Positivism and External Reality", *International Forum* 1, 1(1931)12-16; 2 (1931) 14-19.
1932:*Theory of Electricity and Magnetism*, vol. III of M . Planck, *Introduction to Theoretical Physics*, Mcmillan, London, first edition 1922.

H. Poincaré
1890: *Electricité et Optique*, Carrè et Naud, Paris.

Popper K.
1935: *Logik der Forschung*, Springer, Wien.
1959: *The Logic of Scientific Discovery*, Hutchinson, London; revised edition 1968. New Appendices, 454.
1982: " Quantum theory and the Schism in Physics " in:

Postcript in the Logic of Scientific Discovery, Bartley.
1963: *Conjectures and Refutations. The Growth of Scientific Knowledge*, London.

Raleigh L.
1915: "The Principle of Similitude", *Nature*, No. 2368, 95 (1915) 66-68.

Schrödinger E.
1926 a): "Quantisierung als Eigenwertproblem", *Annalen der Physik* , 79
Erste Mitteilung 79 (1926)   361-376;
Zweite Mitteilung 79 (1926) 489-527;
Dritte Mitteilung 80 (1926) 437-490;
Vierte mitteilung   8I (1926) 109-139.
1926 b): "Über das Verhaltniss der Heisenber-Born-Jordanschen Quantenmechanik zur der meinen", *Annalen der Physik* 79 (1926) 734-756.
1927: "Über den Comptoneffect", *Annalen der Physik* 82 (1927) 257-264.
1928: "La Mécanique des Ondes, Electrons and Photons" *Rapports et Discussions du cinquième Conseil de Physique*, Gauthier-Villars,Paris, 185-213.
Also in: Schrödinger , *Gesammelte Abhandlungen* , Band 3, 185-206.
1929: "Neue Wege in der Physik", *Elektrotechnische Zeitschr*, 50 (1929) I5-16.
1931: "Über Indeterminismus in der Physik - Ist die Naturwissenshaft Milieubedingt?" *Zwei Vorträge zur Kritik der naturwissenshaftlichen Erkenntnis*. (A lecture to the Congress of the Society for the Teaching of Philosophy 16 June , I931); Leipzig I932.
1932:*Über Indeterminismus in der Physik* , Leipzig.
1935: "Die Gegenwärtige Situation in der Quantenmechanik", *Die Naturwissenschaften* , 23 (1935) 807-812 824-828, 844-849.

1950: "What is an elementary particle?", *Endeavor*, Vol IX, N 35, July 1950. Also in : Schrödinger [1984] 456-463.
1951:*Science and Humanism (Physics in our time)*, Cambridge University Press.
1953: "The Meaning of the Wave Mechanics"; in: Various Authors, *Louis DeBroglie, Physicien et Penseur* , Albin Michel.
1958: "Might perhaps Energy be a merely statistical Concept?", *Il Nuovo Cimento*, (10) 9 (1958) 162-170.
1987: *Schrödinger, L'Immagine del Mondo*, Presentazione di B.Bertotti, Torino, 20-36.
1984: *Gesammelte Abhandlungen* , Wieweg & Sohn, Wien, 3 vols.
1984 b): in: Schrödinger [1984] 456-463. Schrödinger's paper was originally published in *Endeavor*, vol IX, No 35, July 1950.

Sommerfeld A.
1935: "Über die Dimensionen der electrodynamische Grossen", *Phys. W.*, 35 (1935) 814-820.
1964: *Electrodynamics*, *(Lectures on Theoretical Physics)* vol. III, Academic Press Boston Mass.

Weber W.
1846: "Elektrodynamische Massbestimmungen uber ein allgemeines Grundgesetz der elektrischen Wirkung, *Annalen der Physik* 73(1848) 193-240. Also in:*Wilhelm Weber's Werke* [1893] 25-214.
1851: "Messungen galvanischen Leitungswiderstande nach einem absoluten Maasse" *Annalen der Physik* 82 (1851) 337-379. Alsoin : *Wilhelm Weber's Werke* [1893] 276-308.
1852: "Elektrodynamische Maasbestimmungen insbesondere Wiederstand Messungen" *Abhandlungen sächs. Ges. Wiss., math-physische Klasse*, Bd.1, Leipzig, I, 199-381. Also in : *Wilhelm Weber's Werke* [1893] 301-465.
1846: " Elektrodynamische Massbestimmungen uber ein all-

gemeines Grundgesetz der elektrischen Wirkung ", *Wilhelm Weber's Werke* [1893] 157.

1848: "Elektrodynamische Massbestimmungen uber ein allgemeines Grundgesetz der elektrischen Wirkung "(Excerpt); a shorter version of the 1846 paper, in: *Annalen*, 73 (1848) 193-240. Also in: *Wilhelm Weber's Werke* [1893] 215-274.

1851: "Messungen galvanischen Leitungswiderstande nach einem absoluten Maasse"*Annalen*, 82 (1851) 337-379.

1856: "Vorwort bei der Übergabe der Abhandlung: Elektrodynamische Maasbestimmungen insbesondere Zurückfürung der Strömintensitäts-Messungen auf Mechanische Maass", in:*Verh. sächs. Ges. Wiss..*, 17(1855) 55-61. Also in:*Wilhelm Weber's Werke* [1893] 591- 596.

1864: " Elektrodynamische Massbestimmungen insbesondere elektrische Schwingungen", in:*Wiilhelm Weber's Werke* [1893] vol. 4, 107-241.

1893:*Wilhelm Weber's Werke*, 5 vol.s, Heinrich Weber, Berlin.

1893: "Galvanismus und Elektrodynamik", *Wilhelm Weber's Werke* [1893] vol. 3, (part l) 276-308.

Weber W. & Koholraush R.

1856: " Über die Elektrizitätsmenge welche by galvanischen Ströme durch der Querschnit der Kette fliesst", *Annalen* 99 (1856) 10-25. Also in :*Wilhelm Weber's Werke* [1893] 597 - 608.

1857: "Elektrodynamische Maasbestimmungen insbesondere Zurückfürung der Strömintensituats-Messungen auf Mecha nische Maass". Submitted by Weber to the Saxon Society in 1855; *Abhandl. sächs. Ges. Wiss.* 3 (1857) 221-90.

BIBLIOGRAPHY
SECONDARY SOURCES

Arendt Hannah,
    1958: *The Human Condition,,* University of Chicago Press.
    1988:*Vita Attiva, La condizione umana*, (Ital. transl.)
    Bompiani, Milano.

Arris R. et al.(eds.),
    1983: *Springs of Scientific Creativity*, Minneapolis.

Baird D., Hughes R.I.G. & Nordmann A.(eds.)
    1998:*Heinrich Hertz: Classical Physicist, Modern Philosopher,* Kluwer Academic Publ.

Barone F.
    1985: *Immagini Filosofiche della Scienza*, Laterza.

Barut A.O.,
    1987: *Annalen der Physik*, Schrödinger Issue, 1987.
    1987: *Foundation of Physics*, Schrödinger Issue,1987.

Battimelli G., Ianniello M.G. (edits),
    1993: *Proceedings of the International Symposium on Ludwig Boltzmann (Rome, 9-11 February 1989),*Verlag der Osterreichischen Akademie der Wissenschaften.

Bitbol M.
    1996: *Schrödinger's Philosophy of Quantum Mechanics*,

Kluwer Academic Publ.

Bergia S.
1988: "Who discovered the Bose-Einstein Statistics ?", in: Doncel, Hermann, Michel (Ed.), *Symmetry in Physics (1690-1890)*, Sevie de Publ. UAB.
1991: "Attempts at Unified Field Theories (1918-1955): Alle ged Failure and Intrinsic Validation/Refutation Criteria", Earmann, Janssen, Norton (ed.), *Einstein Studies Series*, vol.4.
1993: "The fate of Weil's unified theory of 1918" in: F.Bevilacqua (ed.) *First European Physics Society Conference on History of Physics in Europe in the 19th and 20th Centuries,* Ed. Compositori Bologna, 185-193.

Bevilacqua,F.
1984: "H.Hertz Experiments and the Shift towards Contiguous Propagation in the Early Nineties", *Rivista di Storia della Scienza*, Vol 1, n.2 (1984) 239-256.

Bertotti B.(ed.)
1987: *Erwin Schrödinger, L'immagine del Mondo*, Borinhieri, Torino.

Bertotti B.
Presentazione; Bertotti [1987].
1985: *Studies in History and Philosophy of Science* 16 (1985) 83-100.

Bromberg J.
1967: "Maxwell's Displacement Current and his Theory of Light", *Archive for History of Exact Sciences* (1967) 219-234.

Buckingham E.
: 1911: "On Physically Similar Systems: Illustrations of the Use of Dimensional Equations", *The Physical Review,* vol. 4 (1911) 345-376.

Buchwald Jed. Z.,
: 1985: *From Maxwell to Microphysics, Aspects of the Electromagnetic Theory in the Last Quarter of the Nineteenth Century*, The University of Chicago Press.
: 1994: *The Creation of Scientific Effects. Heinrich Hertz and Electric Waves.* The University of Chicago Press.

Bunge M.(Edit.)
: 1967: *Quantum Theory and Reality*, Springe, N.Y., Berlin & Heidelberg.

Bunge M.
: 1973: *Philosophy of Physics,* Reidel, Dordrecht.

Butts R.S.(Ed)
: 1986: *Kant's Philosophy of Physical Science,* Reidel,Dordrecht.

Cappelletti V.
: 1976: *Opere di Hermann Von Helmholtz*, Utet.

Cattaneo G. & Rossi A.(eds.)
: 1991: *I Fondamenti della Meccanica Quantistica. Analisi Storica e Problemi Aperti*, Editel, Commenda di Rende (Italy).

Carneiro Fernado L.
: 1993: *Anàlise Dimensional e Teoria de Semelhanca e dos Modelos Fisicos*, Editora UFRJ, Rio de Janeiro.

Cassirer E.
>1923: *Substance and Function and Einstein's Theory of Relativity,* The Open Court Publ. Co., Chicago.
>1950: *The Problem of Knowledge, Philosophy of Science and History of Science since Hegel,* Yale University Press.
>1953: *Substance and function and Einstein's Theory of Relativity,* Dover reprint.

Cattaneo G, Rossi A.(edits),
>1991: *I Fondamenti della Meccanica Quantistica. Analisi Storica e problemi aperti*; Proceedings of the Camerino Congress, Camerino, Editel, Commenta di Rende (Italy).

Cazenobe J.
>1980:"Comment Hertz a-t-il eu l'idée des ondes Hertziennes?", *Revue de Synthèse,* 111, .89-I00, 345-373.

Chevalley C.
>1989 a: "Histoire et Philosophie de la Mecanique Quantique. Traveaux recent" in: *Revue de Synthese,* IV, 3-4.
>1989 b: "De Bohr et Von Neumann à Kant; L'Ecole allemand de logique quantique", in: *L'Age de la Science,* .2, *Epistemologie,* O.Jacobs
>1991: *Annexes,* in: Niels Bohr, *Physique atomique et con naissance humaine,* Gallimard, 309-627.
>1993: "Niels Bohr's Words and the Atlantis of Kantianism", in: J.Faye and H.Folse (eds.) *Niels Bohr and Contemporary Philosophy*, Reidel.
>I985: "Complementarieté et langage dans l'interprétation de Copenhagen", Revue *d'Histoire des Sciences; Bohr et la Complementarieté,* 38-344, 25I-292.

Cohen R.S.& Helkana Y. (eds.)
>1977: *Hermann von Helmholtz, Epistemological Writings,*

Boston Studies in the Phiosophy of Science, vol. XXXVII, Reidel.

Cohen R.
    1956:."Hertz's Philosophy of Science, an Introductory Essay", in:Hertz [1956] I-XX.

Crowe M.J. ,
    1967: *A History of Vector Analysis*, University of Notre Dame Press, Notre Dame.

D'Agostino S.
    1968 a): "Il pensiero scientifico di Maxwell e lo sviluppo della teoria del campo elettromagnetico nella memoria "On Faraday's Lines of Force", *Scientia* , 103, 1-11.
    1968 b): "I Vortici dell'Etere nella Teoria del Campo Elettromgnetico di Maxwell", *Physis*, Anno X, Fasc.3,188-202.
    1975 : "Hertz's Researches on Electromagnetic Waves" *Historical Studies in the Physical Sciences*, Vol. 6, 261-323.
    1978: "Experiment and theory in Maxwell's works", *Scientia*, 113, 453-468.
    1980: "Weber and Maxwell on the Discovery of the Velocity of light in 19th Century Electrodynamics", in: M.D. Grmek, R.S. Cohen, G. Cimino (eds), *On Scientific Discovery*, Reidel, Dordrecht , 281-293.
    1986: "Maxwell's Dimensional Approach to the Velocity of Light",*Centaurus*, 29, 178-204.
    1989: " Pourquoi Hertz et non pas Maxwell, a-t-il- découvert les ondes électrique ? " *Centaurus*, 32, 66-76.
    1990: "Boltzmann and Hertz on the Bild-Conception of Physical Theory, *History of Science, 28,* 380-398.
    1993 a): "Hertz's Researches and their Place in Nineteenth Century Theoretical Physics", *Centaurus*, 36, 46-82.
    1993 b): "A Consideration of the Rise of Theoretical Physics in Europe and of its Interaction with the Philosophical Tradition", in *First European Physical Society Conference on*

History of Physics in Europe in the Nineteenth and Twentieth Centuries, Ed.Compositori Bologna.
1996 "Absolute Systems of Unites and Dimensions of Physical Quantities: a link between Weber's Electrodynamics and Maxwell's Electromagnetic Theory of Light", *Physis,* vol. 33, Nuova Serie, Fasc.1-3, 5-51.

Darrigol O.
1993:"The Electrodynamic Revolution in Germany as Documented by Early German Exposition of Maxwell's Theory", *Archive for History of Exact Scinces*, 45, 189-280.

Doncel M. G.
1991: "On the process of Hertz's Conversion to Hertzian Waves", *Archive for History of Exact Sciences,* vol 43, n.I 1-27.

Dorling J.
1987: "Schrödinger's Original Interpretation of the Schrödinger Equation: a Rescue Attempt"; in: Kilmister [1987] 16-40.

Duhem Pierre,
1902: *Les Théories électriques de J.C. Maxwell*, Paris.

Elkana Y.
1970: "Helmholtz' 'Kraft': An Illustration of Concepts in Flux", *Historical Studies in the Physical Sciences*, 2, 263-289.
1974: "Boltzmann's Scientific Program and its Alternatives", in: Elkana(ed.),*The Interaction between Science and Philosophy*, Atlantic City N.J., 243-274.

Everitt C.W.F.
1975: *James Clerk Maxwell (Physicists and Natural Philosophers)* C. Scribner's Sons, New York.

Fadner W.L.
   1985:"Theoretical Support for the Generalized Correspondence Principle" *American Journal of Physics*, 53(9)829.

Fasol-Boltzmann I.M.(ed.)
   1990:*Ludwig Boltzmann Principien der Naturfilosofi (Lectures on Natural Philosophy 1903-1906)*, Springer-Verlag, Berlin, N.Y.

Feynmann R. P., Leighton R. B., Sands M.
   1965: *The Feynman Lectures on Physics,* Addyson-Wesley.

Fine A.
   1986: *The shaky Game : Einstein Realism and the Quantum Theory*, The University of Chicago Press.

Freundlich Y.
   1978: "In Defence of Copenhagenism", *Studies in History and Philosophy of* Science, 9, 151-179.

Gillispie G.C.(ed.)
   1972: *Dictionary of Scientific Biographies*, C. Scribner's Sons, New York, vol. 5.

Harmann P.M.(ed.)
   1995: *The Scientfic Papers and Letters of James Clerk Maxwell*, Vol.2:1862-1873; Cambridge University Press.

Heimann P.N.
   1970 "Maxwell and the Modes of Consistent Representation" *Archive for History og Exact Sciences,* 6, 171-213.

Hendry J.
>1984: *The Creation of Quantum Mechanics and the Bohr-Pauli Dialogue*, Reidel, N.Y.

Holton G.
>1973: *Thematic Origins of Scientific Thought. Kepler to Einstein,* Harvard University Press.
>1988: *Thematic Origins of Scientific Thought. Kepler to Einstein,* (Revised Edition) Harvard Uniersity Press.

Honner J.
>1982: "The Transcendental Philosophy of Niels Bohr", *Studies in Hisory and Philosophy of Science*, 13, 1-29.

Hoppe E.
>1928: *Histoire de la Phisique*, Paris.

Howard D.
>1988: "Einstein and *Eindeutigkkeit:* a Neglected Theme in the Philosophical Background to General Relativity", in: Eisestaedt J., Kox A.J.,(eds.),*Studies in the History of General Relativity*, Birkhäuser, 1992; 154-243.
>1990:"Nicht sein kann was nicht sein darf", in: Miller A.(ed.) *Sixty-Two Years of Uncertainty; Historical Philosophical and Physical Inquiries into the Foundations of Quantum Mechanics"*, Plenum Press, 61-112.

Jammer M.
>1966: *The Conceptual Development of Quantum Mechanics,* Mc-Grow Hill.
>1974: *The Philosophy of Quantum Mechanics*, Wiley & Sons.

Janik A., Toulmin S.
>1973: *Wittgenstein's Vienna,* New York.

Janssen M
 1992: "H.A. Lorentz's Attempt to give a Coordinate-Free Formulation of the General Theory of Relativity", in: J.Eisenstaedt, A.J. Kox (eds.), *Studies in the History of General Relativity*, Birkhöuser, 334-365.

Jungnickel C.& McCormmach R.,
 1986: *Intellectual Mastery of Nature. Theoretical Physics from Ohm to Einstein,* 2 vols., The University of Chicago Press; *The Torch of mathematics 1800-1870*, Vol I. *The Now Mighty Theoretical Physics, 1870 -1925*, Vol.II.

Kilmister C.W. (ed.)
 1987: *Schrödinger, Centenary Celebration of a Polymath*, Cambridge University Press.

Kaila Eino
 1950:"Zur Metatheorie der Quantenmekanik", *Acta Philosophica Fennica*, 5, 3-136.

Kalckar J.
 1985: Introduction, NBCW.

Klein M. J.
 1973: "The Development of Boltzmann's statistical Ideas", in: E.G.D. Cohen, W. Thirring (eds.), *The Boltzmann Equation, Theory and Applications*, Berlin, 53-99.

Königsberger L.
 1965: *Hermann von Helmholtz,* (English translation), A. Welby, N.Y.

Krajevski W.
 1977: *Correspondence Principle and Growth of Science*, Reidel.

Kuhn T.S.
 1978: *Black-Body Theory and the Quantum Discontinuity 1894-1912,* Clarendon Press at the Oxford University Press- New York.

Landau L.D., Lifshitz E.M.
 1958: *Quantum Mechanics, Non Relativistic Theory*, Pergamon Press.

Levi-Civita T.
 1917:"Nozione di Parallelismo in una varietà qualunque", *Rendiconti del Circolo Matematico di Palermo*, 42, 137-205.

Maltese G., Orlando L.
 1994: "La condizione di rigidità in Relatività Ristretta e la genesi della Relatività Generale", forthcoming.

Meyenn (Von) Karl ,
 1981: "Boltzmann als kritiker und Rezensent", in: Sexl & Blackmore [1982] 97-127.

Miller A.I.
 1984: *Imagery in Scientific Thought*, Birkhauser, Boston.
 1996: *Insights of Genius; Imagery and Creativity in Science and Art,* Copernicus, Springer Verlag .

Miller A., Hendry J.
 1984: *The Creation of Quantum Mechanics and the Bohr- Pauli Dialogue*, Reidel, N.Y.

Mulligan J F.
   1987: "The influence of H.von Helmholtz on Hertz's contributions to physics", *American Journal of Physics* 55(8), 711-719.

NBCW
   1976-1986: Rüdinger E. (general ed.), *Niels Bohr.Collected Works*, North Holland, 9 Vols.

Stolsenburg (ed),
 1984: *The Emergence of Quantum Mechanics 1924-1926*, NBCW, vol. 5.

Kalchar (ed)
1985: *Foundation of Quantum Physics (1926-1932)*, NBCW, vol.6.

Niven W.D.
   1890: Preface to *The Scientific Papers of James Clark Maxwell*, in: Maxwell [1954] IX- XXIX.

Norton J.
   1984: "How Einstein Found his Field Equations: 1912-1915" *Historical Studies Physical Sciences*, 14(2), 253.
   1985: "What was Eintein's Principle of Equivalence?", *Studies in History and Philosophy of Science*, 16, 203-246.

O'Hara and Pricha W.
   1990: *Hertz and the Maxwellians*, P.Peregrinus.

O'Rahilly A.
   1938: *Electromagnetic Theory, A Critical Examination of Fundametals*, Longmans.
   1965: *Electromagnetic Theory, A Critical Examination of*

Fundametals, Dover repr., 2 vols.

.Pais A.
1982: *Subtle is the Lord. The Science and the Life of Albert Einstein*; Clarendon Press, Oxford.
1991: *Niels Bohr's Times (in Physics, Philosophy, and Polity)*; Clarendon Press, Oxford

Panofsky W. K. H., Phillips M.
1955:.*Classical Electricity and Magnetism*; Addison-Wesley.

Park J.I., Margenau
1968: "Simoultaneous Measurability in Quantum Theory", *International Journal of Theoetical Physics*, 1, 211-283.

Park J.l.
1968: "Nature of Quantum States", A*merican Journal of Physics,* 211-226.

Pouthas J., Oms J.
1981: "L. Boltzmann and the Second Law of the Theory of Heat", presented at the International Conference on L.Boltzmann, Vienna.

Radnitsky G. , Anderson G. (eds.)
1979 *The Structure and Development of Science,* Boston SFS vol. 59, Reidel.

Raveau C.
1894: *L'eclarage èlectrique*, T.1vol 40.
1895: *L'eclarage èlectrique*, T.11.vol.60.

Raileigh L.
    1915: "The Principle of Similitude", *Nature,* no. 2368, 95, 66-68.

Rosenfeld L.
    1956: "The Velocity of Light and the Evolution of Electrodynamics", Il Nuovo Cimento, Supplemento al vol. 4, s. 10, 5, 1630-1667.

Rozental S.(ed.)
    1967: *Niels Bohr; His Life and Work*, Wiley.

Rüdinger E.(gen. ed.)
    1976-1986: *,Niels Bohr. Collected Works,* North Holland, 9 Vols.(shortened : NBCW)

Rüger A.
    1988: "Atomism from Cosmology: Erwin Schrödinger.'s Work on Wave Mechanics and Space-time Structure", *Historical Studies in the Phisical Science* 18:2, 377-401.

Sexl Roman, John Blackmore (ed.)
    1982: *L. Boltzmann Gesamtausgabe* vol 8, Ausgewaehlte Abhandlungen der Internationalen Tagung Wien 1981, Graz 1982.

Schilpp Arthur (ed.)
    1951: *Albert Einstein Philosopher Scientist,* Harper Torch Books, 2 vols..

Schaffer Simon
    1994: "Accurate Measurements is an English Science", in M.N.Wise (ed.), *The value of Precision*, Princeton University Press.

Siegel D.M.
    1991: *Innovations in Maxwell's Electromagnetic Theory. Molecular Vortices, Displacement Current and Light*, Cambridge University Press.

Simpson T.K.
    1966:"Maxwell and the Direct Experimental Test of His Electromagnetic Theory, *Isis*, 57, 411-431.

Smith Crosbie, Wise Northon,
    1989: *Energy and Empire, A biographical study of Lord Kelvin*, Cambridge University Press.

Stolzenburg K.(ed.)
    1984: *Niels Bohr Collected Papers*, NBCW[1976-1986] vol.5

Susskind C.
    1964: "Observations of Electromagnetic Wave Radiation before Hertz", *Isis*, 55, 32-42.

Tagliaferri G.
    1985: *Storia della Fisica Quantistica*, Angeli.

Wessels L.
    1983: "Erwin Schrödinger and the descriptive tradition", in: Arris R.et al.[1983] 254-278.

Whittaker E.,
    1951: *History of the theories of Aether and Electricity*, Nelson, 2 vols.

Wiederkehr K.H.
    1967: *Wilhelm Edward Weber, Erforscher der Wellenbewegung und der Elektricität*, Wissenschaftliche

Verlagsges., Stuttgart.

Wise Norton ,
    1992: "The Maxwell Literature and British Dynamical Theory", *Historical Studies in the Physical Sciences*, vol. 13, Part 1, 175-206.

Woodruff A.E.
    1962: "Action at a Distance in Nineteenth Century Electrodynamics", *Isis*, 53, 439-459.

Vizgin Vladimir P.
    1986: "Einstein.,Hilbert, and Weyl: the Genesis of the Geometrical Unified Field Theory Programm" in: *Einstein and the History of General Relativity*; Birhäuser; 300-314.

Zatkis
    1964: *American Journal of Physics* 32, 898.

INDEX

Ampère XI, XIII, XV, 3, 7, 8, 9, 10, 11, 13-24, 25, 34, 35, 36, 59, 72, 111,113,127,130, 204, 311, 312, 314, 351
Atom 83, 84, 255-258, 276, 283, 284, 290, 321, 343, 348
Atoms 86, 192, 204, 205, 215, 259, 276, 283, 284, 285, 343, 349

Bergia 213, 338, 341, 364
Bertotti 311
Bild XIII, 110, 187, 188, 195, 197, 199, 201, 202, 203, 204, 205, 206, 207, 291, 295, 305, 334, 337, 338, 349, 350, 367
Biot-Savart 19
Bohr XVII, 219, 220, 221, 224, 251, 253-271, 273, 274, 276, 285, 286, 290, 291, 304, 306, 307, 308, 324, 341, 342, 343, 344, 345, 346, 348, 351, 352, 366, 370, 372-376
Boltzmann XI, XIII, XIV, 79, 93, 97, 98, 110, 130, 176, 177, 201, 202, 203, 204, 205, 206, 207, 208, 209, 210, 211, 212, 213, 214, 215, 216, 219, 222, 303, 305, 311, 320, 321, 322, 323, 333, 334, 336, 337, 338, 349, 346, 350, 351, 352, 363, 367, 368, 369, 371, 372, 374, 375,
Born 241, 273, 277, 340, 342, 344, 346, 352, 359
Bridgman 239, 240, 352
Bromberg 315, 316, 326, 364
Buchwald 74, 82, 319, 320, 323, 324, 326, 327, 329, 331, 332, 334, 365
Buckingham 71, 318, 365
Bunge 338, 339, 344, 345

Cassirer XIV, 106, 287, 311, 314, 324, 325, 366
Cauchy XI
Chevalley 324, 336, 341, 344, 349, 350, 366

D'Agostino XVIII, 313, 315, 317, 318, 319, 320, 322, 323, 325, 326, 329, 330, 332, 334, 336, 340
De Broglie 213, 265, 278, 279, 348, 354
Doncel 327, 331, 364, 368
Dorling 350
Duhem 35, 130, 225, 314, 323, 327, 368
Dynamical Theory 56, 58, 60, 80, 82, 84, 201, 321, 357, 377

Einstein XII, XIII, XVI, 6, 7, 73, 75, 106, 213, 214, 224-227, 229-233, 235, 236, 237, 239-251, 252, 255, 261, 264, 265, 269, 270, 271, 274, 276, 280, 291, 301, 303, 304, 305, 307, 308, 311, 319, 323, 338-341, 342, 347, 348, 350-353, 354, 364, 366, 369, 370, 371, 373, 374, 375, 377
Electrodynamics XI, XII, XVI, 3, 7, 8, 19, 21, 22, 26, 29, 32, 36, 37, 39, 42, 43, 45, 49, 70, 74,79, 82, 95, 97, 109, 111, 113, 114, 115, 116, 123, 127, 130, 135, 138, 142, 159, 161, 167, 168, 169, 170, 174, 175, 177, 180, 182, 184, 185, 187, 190, 191, 196, 209, 320, 326, 334, 344, 360, 368, 375, 377
Electromagnetism XII, XIII, 7, 36, 45, 70, 71, 82, 116, 161, 330, 331
Elkana 326, 340, 368

Faraday 7, 15, 16, 18, 21, 22, 33, 34, 49, 52, 54, 78, 100, 111, 117, 119, 121, 127, 132, 133, 148, 159, 167, 167, 174, 175, 176, 178, 179, 183, 332, 333, 357, 367
Fine 338, 341, 346, 347, 350
Fourier XI, 16, 39, 40, 255, 257, 311,

315, 316, 352

Gauss XV, 3, 7, 17, 18, 19, 26, 36, 39, 42, 56, 57, 111, 312, 314, 319, 354
General Relativity (GR) XIV, XVI, 229, 224-227, 2430, 233, 234, 235, 239-242, 244, 245, 247, 248, 249, 252, 276, 323, 339, 341, 349, 370, 371
Goethe 90, 92, 102, 204
Green XI, 48,116

Harmann 321, 322, 326
Heaviside 119, 120, 156, 181, 329
Heimann 316, 369
Helmholtz XIV, XV, XVI, 6, 7, 79, 83, 87-98, 99, 102, 103, 105, 109, 113-116, 122-138, 139, 142, 144, 148, 149, 150, 155, 156, 161, 162-165, 169, 175, 176, 178, 179, 181-185, 192, 195, 197, 198, 219, 303, 311-318, 322, 324, 325-328, 331, 332, 334, 336, 355, 365, 366, 368, 371, 373
Hertz XI, XIII, XIV, XVI, 3, 6, 7, 46, 79, 93- 101, 103, 104, 109, 110, 114, 116, 119-123, 127, 130, 131, 133, 134-165, 167, 168-184, 185, 188-199, 201-211, 214, 303, 305, 322-337, 349, 350, 355, 356, 363-366, 367, 368, 373, 376
Holton XII, 250, 260, 314, 323, 341, 342, 343, 353, 354
Hoppe 120, 192, 193, 325, 334, 337, 346
Howard 341, 353

Jammer 265, 271, 311, 338, 340, 342, 343, 344, 345, 346, 370
Jungnickel & McCormmach 311, 312, 325, 333

Kant XV, 87, 88, 89, 91, 92, 93, 96, 196, 206, 207, 208, 210, 247, 286, 308, 322, 334, 337, 347, 349, 359, 365, 366
Kelvin 109, 163, 198, 315, 318, 356, 376
Kirchhoff 32, 92, 109, 112, 115, 116, 120, 121, 122, 125, 135, 163, 192, 203, 204, 215, 256, 314, 356
Kohlrausch 313, 314, 315, 319
Königsberger 327, 328, 336
Kuhn 336, 338

Lagrange 16, 79, 80, 81, 82, 188
Laplace 188
Levi-Civita 241, 244, 245, 356
Lodge 144, 328, 357
Lorenz 121, 122, 174, 178
Lorentz 7, 123, 130, 215, 227, 230, 276, 331, 371

Maxwell XI, XIII, XIV, XV, XVI, 3-7, 11, 16, 31, 32, 36, 37, 40, 41, 42, 43, 45-56, 58- 63, 65-75, 77-88, 98, 100, 101, 109, 115-118, 120, 121-124, 129-132, 138, 139, 148-155, 156, 159, 160, 161, 165, 167, 168, 174, 175, 176, 178-186, 194-198, 201-204, 219, 220, 313-321, 325, 326, 327, 329, 331, 332, 333, 335, 356, 357, 358, 364, 365, 367, 368, 369, 373, 376, 377

McCormmach 311, 312, 325, 327, 371
Miller 251, 252, 341, 345, 370, 372

Newton XIV, XV, 33, 35, 60, 83, 87, 90, 106, 188, 228, 230, 231, 236, 299, 333
Norton 339

Örsted 7, 13, 16

Panofsky & Phillips 319
Planck XIV, XVI, 41, 109, 185, 212, 213, 214, 215, 250, 257, 271, 303, 315, 325, 333, 338, 341, 343, 358
Petruccioli 342
Poincaré XII, 79, 98, 144, 130, 278, 303, 320, 323, 326, 328, 358
Poisson XI, XVI, 16, 123, 125, 129, 130, 144, 162, 163, 165, 176, 182, 183, 198, 228, 230, 231, 311
Popper 221, 253, 254, 270, 342, 343, 345, 358

Rayleigh 256, 343, 359
Reichenbach 246, 247, 345
Riemann 111, 114, 122, 137, 174, 177
Righi 328, 329
Rossi A. 340, 365, 366

Schaffer 318, 375
Schiller 133, 287
Schrödinger XVII, 273-308, 324, 334, 346, 347, 348, 349, 350, 358, 359, 360, 363, 364, 368, 371, 375, 376
Siegel 319
Simpson 325, 376
Sommerfeld 42, 43, 72, 257, 315, 319, 360
Special Relativity (SR) 73, 74, 75, 226, 230, 231, 233, 234, 235, 236, 239, 240, 241, 311
Stokes XIII, 55, 112, 118, 320, 321, 356

Tagliaferri 271, 343, 346, 376
Thomson W. 40, 48, 49, 52, 53, 58, 69, 70, 77, 80, 83, 84, 112, 113, 115, 116, 118, 119, 120, 121, 122, 143, 151, 188, 316, 318, 319, 320, 321, 325, 326, 356

Unified Field Theory (UFT) 244, 245, 246, 248, 250, 251, 252
Vizgin 325, 326, 341, 356

Weber XIII, XV, 3, 4, 5, 7, 17-37, 39- 42, 45-53, 55- 63, 69-76, 79, 81, 83, 109, 111-116, 120, 124, 125, 128, 130, 132, 133, 137, 163, 171, 173, 175, 176, 177, 203, 312-317, 319, 320, 332, 354, 360, 361, 367, 368
Whittaker 312, 313, 314, 325, 326
Wittgenstein 370

Zermelo 211, 214